SpringerBriefs in Environmental Science

SpringerBriefs in Environmental Science present concise summaries of cutting-edge research and practical applications across a wide spectrum of environmental fields, with fast turnaround time to publication. Featuring compact volumes of 50 to 125 pages, the series covers a range of content from professional to academic. Monographs of new material are considered for the SpringerBriefs in Environmental Science series.

Typical topics might include: a timely report of state-of-the-art analytical techniques, a bridge between new research results, as published in journal articles and a contextual literature review, a snapshot of a hot or emerging topic, an in-depth case study or technical example, a presentation of core concepts that students must understand in order to make independent contributions, best practices or protocols to be followed, a series of short case studies/debates highlighting a specific angle.

SpringerBriefs in Environmental Science allow authors to present their ideas and readers to absorb them with minimal time investment. Both solicited and unsolicited manuscripts are considered for publication.

More information about this series at http://www.springer.com/series/8868

Nibedita S. Ray-Bennett

Avoidable Deaths

A Systems Failure Approach to Disaster Risk Management

Springer

Nibedita S. Ray-Bennett
University of Leicester
Leicester
UK

ISSN 2191-5547 ISSN 2191-5555 (electronic)
SpringerBriefs in Environmental Science
ISBN 978-3-319-66950-2 ISBN 978-3-319-66951-9 (eBook)
https://doi.org/10.1007/978-3-319-66951-9

Library of Congress Control Number: 2017951181

Printed on acid-free paper

This Springer imprint is published by Springer Nature
The registered company is Springer International Publishing AG
The registered company address is: Gewerbestrasse 11, 6330 Cham, Switzerland

The original version of the book was revised: Belated correction to change the last sentence in acknowledgement of frontmatter with new content has been incorporated.

'Etegami for Disaster Victims of the Hiroshima Landslide in 2014' © Picture Letter Group at Denbara and Gion West Community Centers, Hiroshima, Japan

It was a struggle to select an image that was able to effectively illustrate avoidable deaths. Whilst interviewing the Director of the Social Welfare Associate Centre in Hiroshima in 2016, I came across this wonderful Etegami in his office. This Etegami is a combination of hand-made postcards made by the survivors of the Hiroshima Landslide which led to 74 deaths in 2014 in Asaketa and Asaminami wards. I was struck by the powerful images that depict everyday objects, nature, fruits and flowers. It is through these images that victims are remembered and missed by their loved ones.

Our mind-sets are changed now. We are feeling comfortable ki [that) how big the disaster is, let the disaster come—we are ready.
(Distict Emergency Officer of Jagatsinghpur, 29 July 2014, Bhubaneswar).

To
My Dearest Baba

Preface

In 2012, I joined the University of Leicester's Civil Safety and Security Unit (CSSU) from Cranfield University. Whilst teaching the Master of Science in Risk, Crisis and Disaster Management at CSSU, I came across the systems approach. I was instantly attracted to its inherent richness as a macro-paradigm. My previous research had focussed on women and at-risk communities. Therefore, I wanted to engage with organisational-focus risk and crisis studies in order to gather new knowledge for my professional development. Also, I wanted to understand the problem of human deaths from an organisational perspective rather than from the perspective of social vulnerability and risk only.

I decided to return to Odisha for this research. This decision was rather personal. My relationship with Odisha goes back to 1999. When Odisha was hit by the Super-Cyclone on 28 October that year, I was reading my Master of Arts in Social Work at the Tata Institute of Social Sciences in Mumbai. In the aftermath of this Super-Cyclone, I volunteered as a relief worker. Under the guidance of the state government, I spent a fortnight in the district of Jagatsinghpur caring for the widows, orphans and elderly impacted by the Super-Cyclone. This experience was instrumental in shaping my life, career and skills. In 2003, I returned to Odisha once again to conduct nine months of fieldwork for my Ph.D. research in the Department of Sociology at Warwick University. This research was funded by the Ford Foundation International Fellowship Programme. Through this research, I was reunited with the women that I had assisted in the aftermath of the Super-Cyclone. I was also able to document their experiences in depth, not only of the Super-Cyclone but also of later disasters, which included the floods in 2001 and 2003 and drought in 2002 (Ray 2006; Ray-Bennett 2009a, b, c; Ray-Bennett 2010). Therefore, based on my knowledge and experience, I thought Odisha could serve as a 'critical case' (Yin 2012) for systems failure because more than 10,000 people died in the Super-Cyclone.

Importantly, the contacts and networks I established with Odisha State Disaster Management Authority and the local NGOs during my Ph.D. fieldwork were still intact. I have kept in touch with them by sharing publications that have come out of this research. The NGO workers, in particular from Bharat Gyan Vigyan Samity

(BGVS) and Action Aid (AA), became close friends. These well-established contacts and networks prompted me to select Odisha. Furthermore, an opportunity to return to Odisha after eleven years was too tempting to pass up.

I wrote this book at a critical juncture when systems thinking is moribund. Yet, I feel strongly that systems thinking is needed more than ever before in disaster risk reduction studies. I hope that my book is able to generate some interest in this regard. Systems thinking is not an antiquity or something of the past, rather an extremely powerful meta-concept that can contribute to our current practices and change our thoughts. It has certainly changed mine.

Leicester, UK Nibedita S. Ray-Bennett

References

Ray NS (2006) Caste, class and gender: Multiple disasters and women-headed households in an Oriya village. PhD Thesis, Department of Sociology, University of Warwick, UK

Ray-Bennett NS (2009a) Caste, class and gender in multiple disasters: The experiences of women-headed households in an Oriya village. VDM Verlag, Saarbrucken

Ray-Bennett NS (2009b) Multiple disasters and policy responses in pre and post-independence Orissa, India. Disasters 33(2):275–290. doi: 10.1111/j.1467-7717.2008.01074.x

Ray-Bennett NS (2009c) The influence of caste, class and gender in surviving multiple disasters in Orissa, India. Environ. Hazards 8(1):5–22

Ray-Bennett NS (2010) Multiple disasters and micro-credit: Can it reduce women's vulnerability?: The case of Tarasahi in Orissa, India. Disasters 34(1):240–260. doi:10.1111/j.1467-7717.2009.01127.x

Yin RK (2012) Applications of case study research, 3rd edn. Sage, London

Acknowledgements

This book was many years in the making. The idea for this book was first conceived when I was an ESRC Research Associate at Northumbria University (2007–2009). In 2010, I returned to Warwick University. After a short stay at Warwick, I moved to Cranfield University. In 2012, I joined the Civil Safety and Security Unit (CSSU) at the University of Leicester. It was at Leicester that this book became a realistic goal.

In my journey through these Universities, I met some gifted intellectuals who helped me directly and indirectly in shaping this book. I would like to thank each of them. Professor Andrew Collins, Professor Keiko Ikeda, Dr. Hideyuki Shiroshita and Dr. Supriya Akerkar at Northumbria University; Dr. Alice Szczepanikova, Professor Christina Hughes and Professor Duncan Shaw at Warwick University; and Dr. Epimanondas Koronis at Cranfield University. I have received so much from these colleagues, perhaps more than I deserve.

At Leicester, I would like to thank Dr. Simon Bennett and my students for motivating me to write this book. I owe special thanks to Dr. Tony Masys for introducing me to the theories of systems failure. Thanks are also due to Dr. Paul Lawrence for his comments on the book. I am also heavily indebted to my mentor Professor Peter Jackson for helping me to organise my thoughts, reading the first draft of all the chapters, and patiently listening to all my worries and stress of completing this project. Thanks Peter.

I would like to thank the United Nations Development Programme Office in Bhubaneswar and the Government of Odisha, including the Odisha State Disaster Management Authority, Special Relief Organisation, Revenue and Disaster Management Department, Indian Meteorology Department and the District Emergency Officers for being wonderful hosts and also for taking part in this research. I also owe thanks to Mr. Arabinda Ray for helping me to review the relevant reports and records at the Odisha State Disaster Management Authority.

I would like to thank the reviewers for their constructive comments. I would like to thank Mariama San for seeking the permission to use the Etigami for Disaster

Victims from the Picture Letter Group at Denbara and Gion West Community Centers.

Thanks are also due to my Research Assistants: Dr. Lopa Mudra Saxena for transcribing the interviews, Mr. Puskar Jena for collecting death records, and Mr. Bede Wilson for helping me to analyse the death data. Special thanks are due to Bede in particular for his excellent proofreading and formatting work. I would like to thank Mrs. Denise Corsel for helping me out in the last leg of this book. She has been excellent in tidying up the references, frontispiece, index and formatting. I have learnt tremendously while working with my Research Assistants. Thanks to all of you.

I would like to thank my funders for this book project. Hokkaido University for Field Work Fund; Kansai University for a Visiting Scholar Fellow Fund; University of Leicester for a Professional Development Innovation Fund, College of Social Sciences' Research Development Fund and a Study Leave. It would have been absolutely impossible to complete this project without these funds.

I would also like to thank my family in the UK and India for their support. Thanks to my dear husband, Adam for quietly sharing the joys, stress, pain, anger, frustration and happiness of writing. Special thanks also go to our little Asha for making things easier for me. She took my field visits and long hours of work all in her stride. I hope she reads this book when she is able to. The writing process has been extremely challenging. During my Study Leave for this book (January–July, 2015), I lost three relatives in India whom I have loved dearly. My maternal uncle Chimthu-Mama, cousin Nikhil-Da and my paternal uncle's wife, Jethi. They all died rather prematurely. In the last leg of my book, my father was diagnosed with a terminal illness. My father has given me everything that I wanted in my life: love, laughter, strength, support, courage, motivation, dreams and happiness. Words defy the extent of my gratitude, love and respect for my father. I dedicate this book to him.

About The Book

Avoidable Deaths presents systems failure as an analytical tool to explain why deaths occur in disasters and how they may be reduced. Deaths in disasters are a complex problem because the decisions that save lives during disasters sit across different actors and organisations at different levels—local, national, regional and global. The disaster management system, in this context, is a conglomeration of different professional groupings and actors designed for specific tasks and goals. It is also a system that is highly reliant on technology. Actors within this system adopt different frames of reference. As such, weak forms of organisation between the actors could potentially lead to systems failure. These failures have not been fully examined by risk and vulnerability specialists. Underpinned by the theories of justice, systems, risk and crisis management, the analytical tool of systems failure brings these elements to the forefront. More concretely, this is understood through the problems of coordination, communication and the conflicting world views of the relevant actors and organisations with regard to the generation and dissemination of early warning information. The advantage of this analytical tool is assessed in the context of two disasters: the Super-Cyclone of 1999 and Cyclone Phailin of 2013 in Odisha/India. The systems failure approach demonstrates that deaths in disasters are socio-technical failures, and that it is by rectifying systems failures and promoting systems alignment that deaths can be prevented.

Avoidable Deaths is suitable for students, academics, policy-makers and practitioners interested in disaster risk reduction, human rights, gender, sociology of risk and crisis, sustainable development goals, the Sendai goal one, environmental science, organisation and management studies.

Contents

About the Author

Funded by Ford Foundation's International Fellowship Programme **Dr. Nibedita S. Ray-Bennett** completed her Ph.D. in Sociology from Warwick University. Currently, Nibedita is Lecturer in Risk Management in the School of Business's Civil Safety and Security Unit, University of Leicester. Previously, Nibedita has worked with the Ministry of Rural Development and CAPART in India and Warwick, Northumbria and Cranfield Universities in the UK. Her research interests have been in the areas of sociology of risk, crisis and disaster risk reduction, health security, reproductive health, micro-credit, gender, soft systems thinking and critical reflective practices. Most recently, her research explores the role that earth observation can play in disaster risk management. Her research projects have taken her to India, Bangladesh and Japan. She is the author of *Caste, Class and Gender in Multiple Disasters* (VDM Verlag, 2009).

Nibedita is a Research Affiliate at: Northumbria University's Disaster and Development Network (formerly known as Disaster and Development Centre); the National Centre for Earth Observation; University of Leicester's Centre for Climate Change and Landscape Research; and University College London's ESRC funded RELIEF Centre Global Associates International Network (GAIN). Nibedita is a Fellow at the Higher Education Academy in the UK.

List of Figures

List of Tables

Chapter 1
Avoidable Deaths in Disasters

Abstract This chapter presents the concept of 'avoidable deaths' conceived from the theories of risk, violence, justice and organisation. Avoidable deaths are preventable deaths due to advancements in disaster management science and weather forecasting systems; increased sophistication in human-built environments, as well as ongoing economic and policy development worldwide. When avoidable deaths continue to happen, this is event violence. Violence is commenced by the relevant actors and organisations in failing to protect or save lives. Deaths in disasters, in this vein, are a case for violation of justice. In the event of disasters, justice is denied to those women, men and children who would have otherwise lived a long life and an accomplished life. In order to promote justice in disasters, three arguments are put forward. First, human deaths must be identified as a matter of justice; as such they should receive a high priority from the disaster management system by developing a 'goal' to reduce death. Second, problems can be framed, as well as solved within a disaster management system through support-led processes, such as effective INGO, NGO, community and government organisation coordination and communication. Third, it can make room for demands of duty from the actors and organisations involved in protecting lives (Sen, Bull World Health Organ 77 (8):619–623, 1999; Sen, The idea of justice, 2009). This chapter also introduces the case study of Odisha along with the research methods used to conduct the fieldwork between 2013 and 2014.

Keywords Avoidable deaths · Disasters · Disaster management · Goal · Odisha · Justice · Violence

Natural or environmental disasters are understood as hazards which are natural in origin (such as floods, cyclones, tsunami, typhoons, heat wave and lightning), but have the potential to cause human loss and injury. 300,000 human deaths were reported in the Indian Ocean Tsunami in 2004; 15,844 deaths in the Great East Japan (also known as Tohoku) in 2011; 6,000 deaths in the super typhoon Haiyan

---

The original version of this chapter was revised: Belated correction to change a term in chapter abstract has been incorporated.

N.S. Ray-Bennett *Avoidable Deaths*, SpringerBriefs in Environmental Science,
https://doi.org/10.1007/978-3-319-66951-9_1

in 2013 and more than 7,000 deaths in the Nepal earthquake in 2015—to mention a few. However, there is an indication that the number of deaths arising from natural disasters are decreasing (with the exception of 2004) despite an increase in the number of disasters from approximately 90 annually in the 1970s to approximately 450 annually more recently (International Federation of Red Cross and Red Crescent Societies (IFRC) 2013; Plan International 2013; Ray-Bennett et al. 2014a; United Nations International Strategy for Disaster Reduction (UNISDR) 2015a). But the poor and vulnerable continue to die disproportionately in disasters in the developing countries as compared to developed countries (Coppola 2011; Department for Foreign and International Development (DFID) 2013). Between 1980 and 2000, 53% of the deaths due to disasters occurred in poor countries, although these countries accounted for only 11% of the world's 'at-risk' population (Coppola 2011). This merits the question: *why do deaths occur in disasters*?

It is argued here that most of the deaths in disasters are 'avoidable[1]'. Avoidable deaths are preventable deaths due to advancements in disaster management science and weather forecasting systems (DLR 2013; Galtung 1969; Glantz 2009), increased sophistication in human-built environments (Lalonde 2011; UN 2005), as well as ongoing economic and policy development worldwide (Farmer 2004; Sen 2003; UN 2015; UNISDR 2015a, b). Nonetheless, avoidable deaths continue to happen despite these advancements in developed and developing countries in particular. According to Roberts (2008: 9), avoidable deaths are not accidents, "but are instead end points in a causal chain created by human beings". Paul Farmer, a renowned medical anthropologist calls these preventable deaths 'stupid deaths' (Whiteford et al. 2009). When these stupid deaths continue to occur, they become 'violence' (Galtung 1969; Farmer 2004; Rylko-Bauer et al. 2009) or, in the case of disasters, a concept which is called 'event violence[2]'.

Event violence occurs because of pre-existing vulnerabilities that exist in the form of class, gender, caste, race (Bradby 1996; Bradshaw and Fordham 2013; Cockburn 2004; IHRR 2014), and the structures and processes of neo-liberal financial institutions in the developing nations (Roberts 2008). Societal and financial structures and processes often perpetuate existing vulnerabilities (Galtung 1969; Roberts 2008). Cockburn (2004) calls this a 'continuum of violence' which gets magnified during war, military violence or, in the case of this research, during natural disasters. Violence against women and men continues even in the absence of war and during peace (Giles and Hyndman 2004) in varied forms, including under-employment, low wage, wage gap, lack of investment on health, education and disaster management, lack of social protection policies for the poor and vulnerable—to mention a few.

[1]For instance, the Government of Odisha drastically reduced the human casualty in 2013 compared to the Super-Cyclone which had more than 10,000 casualties.

[2]According to Galtung (1969), the father of violence and conflict studies, violence may fall into one of two categories: direct and indirect. Direct violence occurs when an actor with an intention to hurt or harm an individual/s is present; where there is no such actor this is referred to as structural or indirect violence (Galtung 1969, 1985). In the case of disasters, both direct and indirect violence are played out.

Organisations (both governmental and non-governmental) which manage and mitigate disaster risks often contribute to these vulnerabilities either by overlooking pre-existing vulnerabilities or not taking appropriate action to reduce them. This is discussed at length in the next chapter. These pre-existing vulnerabilities exacerbate the impact of disasters and violence by causing human loss, injury and injustice for the deceased.

1.1 Justice

Deaths in disasters are a case for violation of justice. In the event of disasters, justice is denied to those women and men who would have otherwise lived a long life and an accomplished life (Sen 2009). Here the disaster becomes a case of injustice because human deaths could have been prevented by undertaking preventive actions, but the relevant actors and organisations have failed to protect precious lives (Arendt 1970; Farmer 2004). In this light, the lens of event violence offers agency to the deceased rather than victimhood. When human deaths are construed as event violence it offers hope for resolution by suggesting the appropriate preventive measures that might be required. Furthermore, the lens of event violence also empowers every citizen in this world to exercise their right to be saved and protected by the government, irrespective of one's class, caste, race or gender identities. Sen's 'theory of justice' is succinct here to address event violence because it gives a vantage point from which to argue that these deaths are 'intolerable injustices'. Intolerable injustice requires an 'overwhelming priority' without "requiring the search for a consensus on what a perfectly just society would look like" (Sen 2009: 21). Intolerable injustices are the manifestation of human systems failure, and they can be rectified.

The theories of justice are popular amongst feminist studies. Nancy Fraser's theory on politics of recognition and misrecognition is one such example. Inspired by the contemporary moral and political philosophers,[3] Fraser postulated the politics of recognition as a reaction to the global neoliberalism to redress gross mal-distribution and redistribution issues in America and Europe. The politics of recognition gave way to the politics of representation by the third wave of feminists (see Marchand and Parpart 1995; Mohanty 1997). This feminism is not limited to Europe only but in transnational spaces in association with UN agencies and the World Social Forum (Fraser 2001, 2005). They have also problematised woman as a 'point of analysis' by deconstructing 'sisterhood' and universal gender norms and identities that the previous feminist movements have promoted uncritically. Proponents redressed the redistributive and misrecognised issues of those women who are poor and marginalised by their gender and race in the East (with relevance

[3]The provenance of these theories is from two notable philosophers: Kant's distributive justice (morality) and Hegel's recognition of good life (ethics) (Fraser 2005).

in the West too), through the politics of representation (see Fraser 2005; Marchand and Parpart 1995; Visvanathan et al. 1997).

'Gender and disaster' advocates have benefitted much from the politics of recognition, misrecognition and representation (Ray-Bennett 2016a). Visibility of women's deaths in disasters, acknowledgement of women's differential vulnerabilities prior to disasters, absence of women in disaster management practices, and recognition that the disaster experiences are gendered during and after disasters by the national and international actors are some of the outcomes. Likewise, gender mainstreaming policies and programmes promoted by the UN's International Strategy for Disaster Reduction (ISDR) are a testament to the redistributive policies. However, recognition of women and men's deaths in disasters, as well as recognition of these deaths as a case of event violence owing to the failure of disaster management systems is currently amiss. In this context, Sen's (2009) 'theory of justice[4]' is invoked to rectify this gap.

Consequently, how would justice be advanced for human lives in disasters? Three arguments are suggested for this. First, human deaths must be identified as a matter of justice. As mentioned earlier, deaths in disasters can be reduced given the sophistication that has been achieved in science, technology, policy and planning, yet deaths continue to occur. Accordingly, avoidable deaths should receive a high priority in order to address this injustice. One way of prioritising this is to develop a 'goal' to reduce deaths which is discussed in depth later in this chapter. Second, Sen's (2009: ix) theory of justice also clarifies "how we can proceed to address questions of enhancing justice and removing injustice, rather than to offer resolutions of questions about the nature of perfect justice". This approach is empowering in the sense that it enables suggestions of some ways to improve the disaster management system for developing countries in general rather than demanding a complete state of development through transcendental institutions or a just society. As such, problems can be framed as well as solved within the disaster management system.

Third, and most importantly, Sen's (2009: 18) theory of justice is focussed on social realisation "which goes well beyond the organisational picture and includes the lives that people manage or do not manage to live". Social realisations are "assessed in terms of their utilities or happiness [...] [and] the substantive freedoms that people enjoy". This should certainly make organisations and actors *accountable* for what they do, or do not do, to save lives. This is because social realisations are based on a 'capability perspective[5]'—"one that can make room for demands of

[4]Sen's theory of justice is based on the 'alternative theories' of justice spearheaded by Smith, Bentham, Mill, and Wollstonecraft—to mention a few. They differ from the mainstream political philosophers such as Hobbes, Rousseau, Kant, Locke, Rawl—to mention a few—on three grounds: (i) contractarian versus comparative; (ii) arrangement-focussed versus realisation focussed; (iii) just institutions (transcendental institutions) versus different ways in which people live may be led or influenced by institutions and individual's behaviour (for detailed discussion, see Sen 2009).

[5]Capability is the power to do something whether this is related to evacuation prior to a disaster or being able to be well nourished or healthy in everyday life.

duty" (Sen 2009: 19) from the actors and organisations who have failed to protect or save human lives. However, making actors accountable is difficult, particularly considering the poor death reporting systems which are complicated by local context and cultural norms. This is illustrated through the case of Odisha in India in Chap. 4.

1.2 Practical Solutions

Sen's theory of justice is not only to identify 'redressable injustice' but also to suggest practical solutions. Some potential solutions are proposed in this section. To achieve 'avoidable deaths' it is important to understand that not all deaths in disasters are avoidable. Some deaths will continue to occur due to natural causes or illnesses which are pre-existing. For instance, pre-existing health conditions (such as age related disability, heart problem and the like) of an individual can exacerbate causing death during the time of a disaster. Therefore, two types of indicators emerge: avoidable deaths and unavoidable deaths.

Avoidable deaths[6] are understood as deaths beyond the number ten. They are potentially avoidable deaths from the impact of natural disasters (such as cyclone, flood) in the present time, given available knowledge on the nature of these hazards and due to advancements in information technology, human interaction and effective policy interventions. Unavoidable deaths, on the other hand, are understood as anything less than ten deaths. These indicators have both quantitative and qualitative elements and can act as targets to be achieved by the international, national and state disaster management authorities. Indicators both at global and national levels are important milestones or benchmarks for sustainable human development (UN 2015; Wahlström 2015).

Indicators for avoidable deaths are common in the health and development sector. For instance, avoiding maternal mortality (Goal 5) or infant mortality (Goal 4) as suggested by the Millennium Development Goals. These Goals are now encapsulated in Goal 2 (Good Health and Wellbeing) of the Sustainable Development Goals (2015–2030)—successor of the Millennium Development Goals. Accordingly, funds are allocated by the UN and other international and national organisations to achieve this target. WHO (2013: 11) in particular, takes avoidable deaths as an important benchmark to "assess potential for prevention and treatment scale-up and progress towards universal health coverage". WHO extended this principle of avoidable deaths in the floods of Pakistan in 2010, which is explained in the words of the Assistant Director-General of WHO, Dr. Eric Laroche:

> The objective of WHO and our health partners in Pakistan is to reduce avoidable death and illness through a range of life-saving interventions for all people - men, women and children (WHO 2010).

[6]The number of casualties for 'avoidable deaths' is inspired by the definition on 'disasters' provided by the Centre on the Epidemiology of Disaster (CRED).

The availability of resources and appropriate interventions at the right time (disaster climate or beyond) is always a concern to avoid deaths, particularly in developing countries. The development of indicators for avoidable deaths in disasters will have budgetary implications as well as some other implications explained in the words of Murray and Frenk (2000):

> all [avoidable] deaths to be judged in terms of what else could have been done and what would have happened in that event. It requires a judgement of whether we should take the available resources in that country as given, or ask what could have been done if the best available technology had been available. This type of work is certainly beyond the capacity of most countries presently.

1.3 Goal

Disasters are predicted to increase rapidly according to the reports on global warming (Gillies 2014). A majority of meteorologists and climate change experts also believe that the increase in disasters is likely to have an effect on mortality (Lass et al. 2011) and morbidity (WHO 2016). Mortality or deaths in disasters illustrate profound human insecurity (Roberts 2008). It is still rare for the non-economic impacts of the death of a woman or man to be documented; suffice to say that it is a significant tragedy for the family and, in some cases, for the entire rural community in developing nations. Our knowledge of how family members grieve after disasters is still limited. Does it only disrupt the everyday practices of life, or does it have the potential to spawn future crises and disasters at household and community levels (Vollmer 2013)? These questions require further research and examination. It is posited that the UN's 'Sendai Framework for Disaster Risk Reduction 2015–2030' (from now on referred as the Sendai Framework), whose first global target is now 'to reduce disaster mortality', can open up an opportunity to initiate a global and local debates on deaths.

The Sendai Framework is "built on elements which ensure continuity with the work done by states and other stakeholders under the Hyogo Framework for Action 2005–2015" (UN 2015: 5). The Sendai Framework is approved by the 185 UN Member States in the 'Third UNs World Conference on Disaster Risk Reduction', held from 14 to 18 March 2015 in Sendai, Miyagi, Japan (UN 2015). The Sendai Framework is a "15-year, voluntary, non-binding agreement which recognizes that the State has the primary role to reduce disaster risks but that responsibility has to be shared with other stakeholders including local government, the private sector and other stakeholders" (UN 2015).

Much like the Millennium Development Goals, the Sendai Framework has set seven global targets for the national and international actors and organisations. The first global target is to 'reduce global disaster mortality by 2030'. This is welcome news for research because this will lead to allocation of fund, specific research and actions. In this light, this research is timely and relevant.

If reducing mortality is now a global and national goal, this begs the question how such a goal can be translated into practice. It is argued that such a goal can be achieved by adopting a rational decision making model. In management science "the goals are called the objectives or objective functions to be maximised" (Pfeffer 1981: 19). Occasionally, goals are also referred to as preferences based on the world views of the different social actors. The rational model (Pfeffer 1981) demands that goals are consistent. Once a consistent goal is set, the next step is to choose a set of decision making alternatives. Alternatives are determined by the concept of bounded rationality which is the limitations of the responders and organisations involved in terms of information and resources. The bounded rationality is underpinned further by the concept of satisfaction which is determined by the actor's or the responder's level of aspiration and available resources. The next step after sorting the alternatives is to assess the likely outcomes or consequences of the various possible courses of actions. The rational choice model assumes that actions and consequences can be "fully anticipated, albeit with some degree of uncertainty" (Pfeffer 1981: 19). In this model, actors and organisations select the courses of action as well as the alternatives with the ultimate aim to maximise "the achievement of the preferences or goals in objective functions" (Pfeffer 1981: 20).

However, the caveat of this model is that the "assumption of consistency and unity in the goals, information and decision making processes is problematic" (Pfeffer 1981: 20). The real world is messy and political because powerful actors can hinder rational decision making processes due to their vested interests. The other caveat of this model is that it understands that actors and organisations are well equipped to define their goals and alternatives, but it will be noticed in the case of the Super-Cyclone in Odisha that this is often not the case. The head of the state and its bureaucracy failed to rise to the situation with a goal to reduce deaths (Mudur et al. 2013; Ray-Bennett 2009a). On the other hand, when actors and organisations are able to identify their goals and alternatives, the outcome is phenomenal as will be seen in the case of the Cyclone Phailin in 2013. Through these two events the translation of this rational decision model is also understood in Chaps. 3 and 4 using the narratives of the first level responders who personalised the alternatives in the context of bureaucracy with the ultimate aim to maximise the achievement of the goal of 'zero human casualties' in their respective districts. Odisha (renamed from Orissa in 2011), an eastern state of India, is chosen as a case study for this research. In order to contextualise this research, the ensuing sections discuss the socio-economic and the disaster management structures of Odisha.

1.4 Odisha: A Case Study

Odisha is situated in the east of the country, bordered by Madhya Pradesh in the west, Andhra Pradesh in the south, Vihar in the north and West Bengal in the east (see Figs. 1.1 and 1.2). According to the 2011 census, the population of Odisha is at about 41 million, which makes it the 11th most populated state in India (Census of

Fig. 1.1 Political Map of India. (Reproduced from Maps of India 2015)

India 2011). The state makes up about 3.4% of the country's population, a figure which was about 3% in the last census of 2001. The state is spread over an area of about 150,000 km^2. making it the 9th largest state in the country in terms of area. The density of population per sq. km. is about 260, fairly well below the national average. Bhubaneswar is the capital city of Odisha and Odiya is the state language.

Odisha has 30 districts (Census of India 2011) of which 13 are coastal (Samal et al. 2003). The coastal districts are highly prone to cyclones and storm surge. Its coastline adjoins the Bay of Bengal (485 km), which makes it four to five times

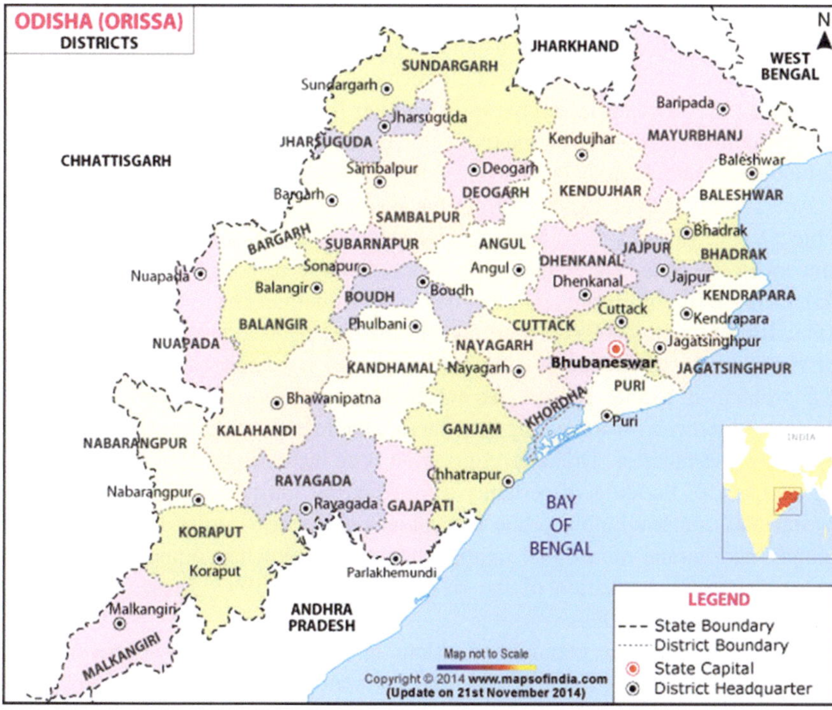

Fig. 1.2 Map of Odisha's Districts. (Reproduced from Maps of India 2014)

more likely to experience storms than it would if it were located in the Arabian Sea (Kalsi 2003; Shiva and Emani 2000). Tropical cyclones[7] from the Bay of Bengal are particularly severe and bring widespread destruction, especially when accompanied by storm surges, high winds and extreme rainfall that results in riverine flooding (Bosher 2005; GoO 2002a, b; Haque et al. 2011). Tropical cyclones are notorious for causing high levels of mortality (Nathan 2009). In 1970 and 1991 cyclones in Bangladesh killed more than 500,000 and almost 140,000 people respectively (Haque et al. 2011). Tropical cyclones also cause high levels of morbidity by affecting general public health, livelihoods, infrastructure, the economy and sociocultural foundations (Haque et al. 2011). The direct impact of a cyclone can cause shortage of food and drinking water, and increase the

[7]The World Meteorological Organization defines a tropical cyclone as "a non-frontal synoptic scale cyclone originating over tropical or subtropical waters with organized convection and definite cyclonic surface wind circulation. More specifically, a storm in the south-east Indian Ocean is cyclonic when the sustained wind speed is more than 33 nautical miles per hour (>62 km/h). […] A storm surge is the difference between the water level under the influence of a disturbance (storm tide) and the normal level that would have been reached in the absence of the meteorological disturbance" (Haque et al. 2011: 150).

transmission risks of infectious diseases, such as diarrhoea, hepatitis, malaria, dengue, pneumonia, eye infections and skin diseases. All of these interrupt livelihood outcomes (Haque et al. 2011). Odisha is generally vulnerable to cyclones from April to May and September to November (GoO 2002a). It is anticipated that the strength and number of major cyclones might be increasing due to higher surface temperatures associated with global warming (Haque et al. 2011).

Apart from cyclonic exposure from the Bay of Bengal, the people of Odisha are subject to the vagaries of complex river systems. Flood[8] is therefore another major concern in the state. The major rivers predominantly responsible for flooding are: Mahanadi, Baitarni, Brahmani, Kathjori and Subarnarekha (Bhatta 1997; GoO 2005; Ray-Bennett 2009a). These are all affected by heavy rainfall in the upper catchments area, which, along with unusual monsoon rainfall, causes them to flood. The problem is further exacerbated when floods coincide with high tides. These block the floodwater from exiting into the sea, having a further devastating effect on low-lying coastal areas. Drainage congestion is an increasingly major contributor to the problem of flooding in certain parts of coastal Odisha. The entire low-lying coastal belt, already highly prone to cyclones and storm surges from the Bay of Bengal, and annual monsoonal rainfall, has these problems compounded through the ever-increasing siltation of the river beds (Bhatta 1997; Bosher 2005; GoO 2005; Ray-Bennett 2009a).

Drought[9] is another perennial problem in the coastal and western areas of Odisha. Approximately 70% of the total cultivated area of the state is prone to drought (GoO 2005). These areas also lack irrigation facilities; only 14% of the state has irrigation provisions, despite the fact that Odisha is an agro-based economy (Pradhan 2003; Ray-Bennett 2009a). The dependence of agriculture on rain means that the slightest variation in rainfall makes drought a constant hazard.

The geographic and climatic conditions of Odisha have led to it experiencing multiple disasters causing recurrent physical and social destruction for people's lives over a sustained period of time. Table 1.1 illustrates the sequence of disasters events affecting Odisha since 1999. Significant rainfall during the monsoon can cause widespread flooding affecting farmland and property, whilst too little rainfall and high temperatures bring drought and heat waves to the state. According to the Orissa Human Development Report, the property lost through disasters in the 1970s was estimated at Indian Rupees 105 crores (Great British Pound 1.3 million), a figure, which increased nearly sevenfold in the 1980s and more than tenfold in the

[8]"Flood is defined as the condition that occurs when water overflows the natural or artificial confines of a stream of other body of water, or accumulates by drainage over low-lying areas. A flood is a temporary inundation of normally dry land with water, [...] overflowing of rivers, precipitation, storm surge, waves, [...]" (Sivakumar 2005: 3).

[9]"Drought is the consequence of a natural reduction in the amount of precipitation over an extended period of time, usually a season or more in length, often associated with other climatic factors (such as high temperatures, high winds and low relative humidity) that can aggravate the severity of the event. Drought is not a purely physical phenomenon, but instead is an interplay between natural water availability and human demands for water supply" (Sivakumar 2005: 3).

Table 1.1 Multiple Disasters in Odisha from 1999 to 2013. (Compiled from Ray-Bennett 2009a and Mohanty 2012)

Year	Multiple disasters
1999	Super-cyclone and flood
2000	Drought
2001	Severe flood
2002	Severe drought
2003	Severe flood, heat wave
2004	Flood and heat wave
2005	Flood and heat wave
2006	Flood and heat wave
2007	Flood and heat wave
2008	Flood and heat wave
2009	Flood and heat wave
2010	Flood and heat wave
2011	Flood and heat wave
2013	Cyclone, heat wave
2014	Flood

1990s (GoO 2005; Ray-Bennett 2009a). This indicates the sheer scale of the threat of natural calamities in a poor state like Odisha, which cause serious imbalances by placing heavy demands on revenue expenditure—that is, the price of restoring assets and meeting the shortfall in income collection due to recurrent crop and property loss (GoO 2005; Ray-Bennett 2009a).

In 2005, the Orissa's Human Development Report declared Odisha as a 'severely food insecure region' due to its susceptibility to natural disasters and their debilitating effect on people's livelihoods in rural areas (GoO 2005; Ray-Bennett 2009a, b). Odisha is one of the poorest states in India. In 1998–1999, Odisha had a very slow rate of agricultural growth (2.38%) (GoO 2005). Although agricultural growth continues to remain low due to a lack of investment, Odisha's industry and service sectors have thrived since 2003. In 2012, Odisha's annual average growth rate was 7.6%, consistent with the national level growth rate (Mohanty 2012). In 1999–2000, approximately 47.15% of people were living below the poverty line,[10] a shockingly high figure compared to the all India average of 26.1% (GoO 2002b; Ghosh et al. 2012). About 83.31% of people still live in the rural areas compared to 16.69% in urban areas (Census of India 2011). The density of population is still the lowest compared to other poor states such as Vihar and Madhya Pradesh (Census of India 2011). Odisha continues to have the highest infant mortality rate in the country (87 per thousand children in 2002), signifying a low status on the human development scale (GoO 2005; Mohanty 2012; Ray-Bennett 2009a, b).

[10]The poverty line in India is defined according to lack of income. In 1993–94, the poverty line was defined as Rs. 205.84 (GBP 2.57) per capita per month in rural areas and Rs. 281.35 (GBP 3.51) in urban areas (Cockroft 2003).

1.5 Disaster Management in Odisha

During the colonial period, the British government undertook certain mitigation measures aimed at attenuating the dangers of multiple disasters. However, these were mostly reactive in approach. From 1850–1939, several committees were set up in response to events such as cyclones, droughts and floods. Important committee documents released during the nineteenth and twentieth centuries were the *Harris Committee Report*, *J.P. Beadle's Report* and *Col. Arthur Cotton's Report* (Bhatta 1997; GoO 1996). The committees frequently recommended structural mitigation measures, as well as some non-structural initiatives, to reduce the impact of droughts and floods, including:

i. an elaborate system of canals;
ii. weirs across the Baitarnai, Brahmani and Mahanadi Rivers;
iii. irrigation channels throughout the deltaic region for navigation;
iv. embankments on all rivers;
v. development of a meteorological department; and
vi. the extension of early warning facilities from Calcutta (currently known as Kolkata) to Orissa ports to signal oncoming cyclones.

Several of these measures were implemented in the coastal provinces, but most were delayed because of a lack of response and budgetary constraints and were only implemented following the Great Famine in Orissa (*Nan-ak-Durvigya*) in 1866 (Bhatta 1997).

The second Indian Famine Commission (of 1898) viewed disaster mitigation as an important aspect of disaster management in predicting famine and alerting the state (Bhatta 1997). The third Indian Famine Commission (of 1899) emphasised speedy relief, the provision of fodder for cattle, prompt remission and suspension of land revenues, swift loan distribution, the introduction of fodder camps, and gratuitous relief for women, children and the destitute. Here, uniquely, the urgency surrounding the immediate needs of vulnerable populations was translated into practical measures (Bhatta 1997; Samal 2003). As a result, famine commissions during the British period were seen as exemplary in addressing for the first time the immediate needs of people at the household level and for looking to prevent large numbers of deaths in famines (Samal 2003). However, their relief responses were extremely short term and ad hoc in nature. Consequently, little was done to tackle poverty or vulnerability, nor indeed to boost people's capacity to cope with multiple disasters (Bhatta 1997; Ray-Bennett 2009a, b).

In the post-independence era, the Government of Odisha inherited the colonial Famine Relief Code which was amended in 1988 and called the Orissa Relief Code (GoO 1996). Orissa Relief Code (from now on referred as The Relief Code) was the only disaster management policy document up until 2005. The Relief Code specified how administrators should identify crisis conditions, how they should respond, and when they should do so during natural disasters (Currie 2000; GoO 1996;

GoI 2007). To operationalise the Relief Code, several actors, departments and organisations are involved in the management and mitigation of disasters in Odisha. A detail of these stakeholders is provided below.

1.6 Disaster Management Organisations

Revenue and Disaster Management Department: 'This Department is responsible for providing immediate relief to the people affected by natural disasters such as floods, droughts, cyclones, hailstorms, earthquakes, fire accidents, etc. It is also responsible for relief, rescue, rehabilitation and restoration work. The Department is headed by the Principal Secretary/Addl. Chief Secretary, who exercises all administrative and financial powers for the Department. Special Relief Commissioner, who is ex-officio Special Secretary to Government, Revenue & Disaster Management Department, is assisted by a group of experienced officers and staff' (OSDMA 2014: 39).

Special Relief Organisation: 'The Special Relief Organisation was created under the Board of Revenue in 1965–66 for relief and rescue operation during and after occurrence of natural disasters. Since its inception, the scope of the Special Relief Organisation has been diversified. Currently, it deals with disaster management which involves responsibility of prevention, mitigation, preparedness, response, relief and rehabilitation related to natural disasters. It coordinates with districts/departments for quick relief and rescue operation, reconstruction and rehabilitation work. It also promotes disaster preparedness at all levels in the state with the assistance of Odisha State Disaster Management Authority (OSDMA). The State Emergency Operation Centre is a latest addition at Rajiv Bhawan in Bhubaneswar with state of art communication net-work. The State Emergency Operation Centre functions throughout the year. In the time of natural disasters, it is open round the clock. The Special Relief Organisation is headed by the Special Relief Commissioner (SRC) who exercises all the administrative and financial powers. He is assisted by a group of experienced officers and staff' (OSDMA 2014: 39).

Odisha State Disaster Management Authority: The Government of Odisha set up this autonomous organisation in the intermediate aftermath of the Super-Cyclone in 1999. It is a non-profit charitable institution, with its headquarters at Bhubaneswar and jurisdiction over the whole state. From 26 August 2000, the Department of Revenue has become the administrative Department for the Odisha State Disaster Mitigation Authority. As such, it has been renamed as Odisha State Disaster Management Authority from Mitigation Authority.

'The Odisha State Disaster Management Authority covers the entire gamut of disaster management including relief, restoration, reconstruction, and preparedness. As part of the preparedness activities it:

- Coordinates with the line departments involved in reconstruction,
- Coordinates with bilateral and multi-lateral aid agencies,

- Coordinates with UN Agencies, international, national and state-level non-governmental organisations (NGOs),
- Networks with similar and relevant organisations for disaster management' (OSDMA 2014: 41).

Post 'Disaster Management Act 2005': On 1 August 2007, the Government of India passed the first 'Disaster Management Act, 2005' which urged for a three-tier institutional structure: disaster management at the national, state and district levels in the form of National Disaster Management Authority, State Disaster Management Authority, and District Disaster Management Authority. National Policy on Disaster Management (NPDM) also specified the roles and responsibilities of various organisations for disaster response. As a result, the Government of Odisha created some additional institutions at the State and District Levels which are described below' (OSDMA 2014: 41).

The *State Disaster Management Authority* was officially launched on 20th October 2010. The Chief Minister of Odisha is the ex-officio Chairperson, the Minister, Revenue and Disaster Management is the Vice Chairman, and the Chief Secretary is the Member-cum-Ex-Officio Chief Executive Officer of the State Disaster Management Authority. Other Members include:

i. Minister, Agriculture;
ii. Minister, Finance;
iii. Development Commissioner/Addl. Development Commissioner;
iv. Secretary, Home Department;
v. Secretary, Department of Water Resources;
vi. Secretary, Agriculture Department; and
vii. Secretary, Revenue and Disaster Management Department.

The Special Relief Commissioner's Office is designated as the secretariat of the State Disaster Management Authority. The Special Relief Commissioner is the additional Chief Executive Officer. The State Disaster Management Authority has the responsibility for laying down policies and plans for disaster management State Executive Committee.

The State Executive Committee was constituted to assist and support the coordination for the State Disaster Management Authority. The Committee consists of:

i. Chief Secretary—Chairperson, Ex-officio
ii. Development Commissioner/Addl. Development Commissioner—Member
iii. Agriculture Production Commissioner—Member
iv. Secretary, Revenue and Disaster Management Department—Member
v. Special Relief Commissioner—Member

The State Executive Committee is responsible for implementing the National Plan and the State Plan. It is also responsible for coordinating and monitoring the management of disasters and also in charge of the State Disaster Response Fund.

District Disaster Management Authority is constituted in all districts to oversee Disaster Management activities at district level. The Collector is the Chairperson of

District Disaster Management Authority whereas the Chairman, Zilla Parishad is the Co-Chairperson. The Superintendent of Police, the Chief District Medical Officer and Executive Engineers in charge of embankments, Assistant District Magistrate in charge of Emergency, Executive Engineer for Rural Development Department, Project Coordinator for District Rural Development Authority, and Deputy Director for Agriculture Department are the Members of District Disaster Management Authorities (OSDMA 2014).

1.6.1 Institutional Arrangements for Disaster Management

'The Department of Revenue and Disaster Management is the administrative department for the management of disasters. The Special Relief Commissioner is in charge of the response phase of disasters, whereas, the Odisha State Disaster Management Authority deals with preparedness and mitigation aspects. The Odisha State Disaster Management Authority provides support to the Special Relief Commissioner during the response phase. At the district level, Collector is the District Relief Officer and Disaster Manager. A block is the lowest unit of relief administration. The Block Development Officer and Tahasildars jointly manage the relief administration at the lowest level. The State level Natural Calamity Committee functions under the chairmanship of the Chief Minister for overall supervision and monitoring at the state level. At the district level, the District Natural Calamity Committee along with the District Disaster Management Authority functions with representation from district level officers and peoples' representatives under the chairmanship of the district Collector for supervision and monitoring. Block Disaster Management Committees (under the chairmanship of the Chairperson, Panchayat Samiti), Gram Panchayat Disaster Management Committees (under the chairmanship of the Sarpanch) and Village level Task Force Committees have been constituted in the programmed areas (16 districts) under the Government of India-United Nations Development Programme (GoI-UNDP) Disaster Risk Management programme (DRM) for day-to-day management of disasters and risk reduction measures' (OSDMA 2014: 43).

Indian Meteorology Department in Odisha: Although Odisha's State Disaster Management Plan does not specify the role and involvement of the Indian Meteorology Department in disaster management,[11] in the context of this research and also to reduce deaths, the role of this Department is vital. The Indian Meteorology Department belongs to the Government of India. The purpose of this Meteorology Department is to provide support to the state departments by generating effective early warning information for future planning and disaster response. An effective response system is vital to save lives because most of the causalities in

[11]This may be due to the fact that early warning systems are still not considered as an integral part of disaster management.

cyclones occur at the last phase of disaster preparedness and early phase of response (Prizza 2007). As such, the role of the Indian Meteorology Department is vital in disaster management. The Indian Meteorology Department's Area Cyclone Warning Centres are located at Chennai, Kolkata and Mumbai. These Centres are enhanced by India's new space-based technology the INSAT Satellite—3D. The INSAT-3D satellite was launched by India's Space Research Organisation and the Indian Meteorology Department from French Guiana on 26 July 2013 (PTI 2014).[12] Cyclone Warning Centres are activated on the receipt of warnings from INSAT-3D and these Centres continue to function even when all other means of communication breakdown by relying on battery powered and the INSAT Satellite information (Srivastava 2009). The Director of the Indian Meteorology Department in Bhubaneswar is supported by five Duty Officers and several research personnel. They work closely with the Indian Meteorology Department in New Delhi.

1.7 Fieldwork

The fieldwork for this research was conducted over a period of one year (August 2013 till August 2014). When the fieldwork began, the purpose of this research was to collect data related to the Super-Cyclone only. Cyclone Phailin hit Odisha on 18 October 2013, which was during the research period. As a result, Phailin became another case study because the Government of Odisha drastically reduced the human casualty compared to the Super-Cyclone.

A Research Assistant was hired in Odisha to review death records and reports from 1999 to 2013 from four sampled districts which were Jagatsinghpur, Cuttack, Puri and Ganjam. Ideally more districts could have been selected but this was limited due to funding constraints. The selection of the sampled districts was both purposive and convenient (Bryman 2001). All these districts experienced the Super-Cyclone and Cyclone Phailin. They were the worst affected districts in comparison to the remaining nine coastal districts. They were also the largest districts in terms of population size. Besides, the research team had good knowledge of at least two districts, Jagatsinghpur and Cuttack, because of their previous work in Odisha. This background knowledge was important when designing the research process. Furthermore, the willingness of the District Emergency Offices to partake in this research was also critical to selecting these districts.

Death reports and records were reviewed by the Research Assistant visiting the District Emergency Offices of Jagatsinghpur, Ganjam, Puri and Cuttack. For the purpose of verification, some death reports and related data were also collected from the offices of Odisha State Disaster Management Authority, the Census Office

---

[12]INSAT-3D is considered to be far superior to its predecessors Kalpana-1 and INSAT-3A in terms of weather monitoring and weather forecasting. The launch of INSAT-3D is a major leap for India because this is a first ever venture from a developing country (Union Science and Technology Minister S. Jaipal Reddy; quoted in PTI 2014).

and the Special Relief Organisation in Bhubaneswar. For further verification of the death data, seven interviews were conducted with the government officials by the author.

She interviewed the Director of the Indian Meteorology Department, Government Representative from the Revenue and Disaster Management Department, Deputy General Manager of Odisha State Disaster Management Authority, Deputy Relief Commissioner of the Special Relief Organisation, State Programme Officer of United Nations Development Programme and three District Emergency Officers of Jagatsinghpur, Puri and Ganjam. For the purpose of anonymity participants are referred by their designations. These interviews were recorded using the software MyMemos. The interviews lasted between one to three hours and the participants spoke largely in English with few words in Oriya and Hindi.

The author also conducted several hours of informal talks over tea and lunch with the Deputy Relief Commissioner of the Special Relief Organisation, Director of the Indian Meteorology Department, Deputy General Manager of the Odisha State Disaster Management Authority and the District Emergency Officer of Jagatsinghpur. The informal talks were not recorded, but a field note diary was maintained because some of these discussions were very animated and informative (Hughes 1994).

Informal interviews were also conducted with one of the Clerks at the District Emergency Office in Puri; the Regional Director of Bharat Gyan Vigyan Samity (BGVS); and a Project Coordinator of ADRA in Bhubaneswar. The latter participants were involved in a rehabilitation project after the Cyclone Phailin in Ganjam district. These talks were informal and were not recorded but the author kept notes on these. These talks were extremely informative and they allowed the author to become familiar with the disaster and developmental activities that are happening in Odisha. As a result, they were useful in building the narrative for this research (Hughes 1994). Records/reports, secondary literature, formal and informal interviews were therefore triangulated to enhance quality of the collected data (Burgess 1991).

References

Arendt H (1970) On violence. HBJ Book, London

Bhatta BB (1997) The natural calamities of Orissa in the 19th century. Commonwealth Publisher, New Delhi

Bosher LS (2005) The divine hierarchy: the social and institutional elements of vulnerability in South India. Unpublished thesis. Middlesex University, London

Bradby H (1996) Introduction. In: Bradby H (ed) Defining violence: understanding the cause and effects of violence. Ashgate Publishing Ltd., Aldershot, pp 1–21

Bradshaw S, Fordham M (2013) Women, girls and disasters: a review for DFID. Department for International Development, London

Burgess RG (1991) In the field: an introduction to field research. Routledge, London

Bryman A (2001) Social research methods. Oxford University Press, Oxford

Census of India (2011) Orissa population census data 2011. Census of India. http://www.census2011.co.in/census/state/orissa.html. Accessed 05 Jan 2016

Cockburn C (2004) The continuum of violence: a gender perspective on war and peace. In: Giles W, Hyndman J (eds) Sites of violence: gender and conflict zones. University of California Press, California

Cockroft J (2003) An overview of chronic poverty in India. Int J Rural Stud 10(2):13–14

Coppola DP (2011) Introduction to international disaster management, 2nd edn. Elsevier Inc., London

Currie B (2000) The politics of hunger in India: a study of democracy, governance and Kalahandi's poverty. Macmillan Press Ltd., London

DFID (2013) Helping developing countries deal with humanitarian emergencies. DFID, London

DLR (2013) Earth observation: discovering, surveying, and understanding our planet. Deutsches Zentrum fur Luft-und Raumfahrt e.V (DLR), Bonn

Farmer P (2004) Pathologies of power: health, human rights, and the new war on the poor. University of California Press, California

Fraser N (2001) Recognition without ethics? Theory Cult Soc 18(2–3):21–42. doi:10.1177/02632760122051760

Fraser N (2005) Mapping the feminist imagination: from redistribution to recognition to representation. Constellations 12(3):295–307. doi:10.1111/j.1351-0487.2005.00418.x

Galtung J (1969) Violence, peace, and peace research. J Peace Res 6(3):167–191. doi:10.1177/002234336900600301

Galtung J (1985) Twenty-five years of peace research: ten challenges and some responses. J Peace Res 22(2):141–158. doi:10.1177/002234338502200205

Giles W, Hyndman J (eds) (2004) Sites of violence: gender and conflict zones. University of California Press, California

Gillies J (2014) Scientists sound alarm on climate change. New York Times, March 18. http://www.nytimes.com/2014/03/18/science/scientists-sound-alarm-on-climate.html?_r=0. Accessed 18 Mar 2014

Ghosh S, Verma HC, Panda DL, Nanda P, Kumar A (2012) Irrigation, agriculture, livelihood and poverty linkages in Odisha. Agric Econ Res Rev 25(1):99–105

Glantz M (2009) Introduction. In: Glantz M (ed) Heads up!: early warning systems for climate, water and weather-related hazards. United Nations University Press, New York, pp 1–28

GoI (2007) Disaster management act 2005. Ministry of Law and Justice, New Delhi

GoO (1996) The Orissa relief code. Revenue Department, Bhubaneswar

GoO (2002a) Managing disasters in Orissa: background, challenges and perspectives. Orissa State Disaster Mitigation Authority, Bhubaneswar

GoO (2002b) Orissa state development report. Nabo Krishna Centre for Development Studies, Bhubaneswar

GoO (2005) State disaster management policy: resolution 4th March 2005. Revenue Department, Bhubaneswar

Haque U, Hashizume M, Kolivras KN, Overgaard HJ, Das B, Yamamoto T (2011) Reduced death rates from cyclones in Bangladesh: what more needs to be done?. Bull World Health Organ 2012(9):150–156. doi:10.2471/BLT.11.088302

Hughes C (1994) From field notes to dissertation: analyzing the stepfamily. In: Bryman A, Burgess RG (eds) Analyzing qualitative data. Routledge, New York, pp 35–46

IFRC (2013) World disasters report 2012: disaster dataset. International Federation of the Red Cross. http://worlddisastersreport.org/en/data/index.html. Accessed 30 Jan 2014

IHRR (2014) Keep the light on: using night lights to identify flood vulnerability. Institute of Hazard, Risk and Resilience. http://www.dur.ac.uk/ihrr/. Accessed 17 Mar 2014

Kalsi SR (2003) Monitoring and forecasting of tropical cyclones and their associated effects. In: Gupta HK (ed) Disaster management. Universities Press, Mumbai

Lalonde C (2011) Managing crisis through organisational development: a conceptual framework. Disasters 35(2):443–464. doi:10.1111/j.1467-7717.2010.01223.x

Lass W, Haas A, Hinkel J, Jaeger C (2011) Avoiding the avoidable: towards a European heat waves risk governance. Int J Disaster Risk Sci 2(1):1–14. doi:10.1007/s13753-011-0001-z

Maps of India (2014) District map of Odisha. Maps of India. http://www.mapsofindia.com/maps/orissa/orissa.htm. Accessed 3 Aug 2016

Maps of India (2015) Political map of India. Maps of India. http://www.mapsofindia.com/maps/india/india-political-map.htm. Accessed 3 Aug 2016

Marchand M, Parpart J (eds) (1995) Feminism/Postmodernism/Development. Routledge, London

Mohanty CT (1997) Under the western eyes: feminist scholarship and colonial discourses. In: Visvanathan N, Duggan L, Nisonoff L, Wiegersma N (eds) The women, gender and development reader. Zed Books, London

Mohanty D (2012) Economic and financial developments in Odisha. RBI Monthly Bull 2011:1629–1636

Mudur GS, Mishra A, Mohanty S (2013) System shows its spine: battle 42 years in the making. The Telegraph, 16 Oct

Murray CJL, Frenk J (2000) Can the WHO Health Report improve health systems performance? A response. World Health Organisation. http://www.who.int/health-systems-performance/docs/articles/sjph_murray.pdf?ua=1. Accessed 21 Jan 2014

Nathan F (2009) Natural disasters, vulnerability and human security: facing global environmental change. Hexagon Ser Hum Environ Secur 4(Part II):103–118. doi:10.1007/978-3-540-68488-6_87

OSDMA (2014) State disaster management plan. Government of Odisha, Bhubaneswar. www.osdma.org/Download/State_Disaster_Management. Accessed 6 July 2016

Pfeffer J (1981) Power in organization. Pitman Publishing Inc., London

Plan International (2013) Because I am a girl: the state of the world's girl 2013. Plan International, Woking

Pradhan RP (2003) Agriculture efficiency of Orissa: problems and prospects. J Rural Dev 22 (2):220–247

Prizza R (2007) The role of coordination in disaster management. In: Pinkowski J (ed) Disaster management handbook. CRC Press, New York

PTI (2014) Advanced weather satellite INSAT-3D becomes operational. New Indian Express, 16 Jan

Ray-Bennett NS (2009a) Caste, class and gender in multiple disasters: the experiences of women-headed households in an Oriya village. VDM Verlag, Saarbrucken

Ray-Bennett NS (2009b) Multiple disasters and policy responses in pre and post-independence Orissa, India. Disasters 33(2):275–290. doi:10.1111/j.1467-7717.2008.01074.x

Ray-Bennett NS, Masys AJ, Shiroshita H, Jackson P (2014a) Hyper-risks in a hyper connected world: a call for critical "reflective response" to develop organisational resilience. Input paper prepared for the global assessment report on disaster risk reduction 2015. UNISDR, Geneva

Ray-Bennett NS (2016) Exploring the meaning of securitisation for "gender and disaster". In: Racioppi L, Rajagopalan S (eds) Women and disasters in South Asia: survival, security and development. Routledge, New York, pp 277–301

Roberts D (2008) Human insecurity: global structures of violence. Zed Books, London

Rylko-Bauer B, Whiteford L, Farmer P (2009) Prologue: coming to terms with global violence and health. In: Rylko-Bauer B, Whiteford L, Farmer P (eds) Global health in times of violence. School for Advanced Research Press, Santa Fe, pp 3–16

Samal KC (2003) Coping strategy of weaker sections (dalits and women) in post super-cyclone Orissa (case study of Ersamma block). Nabo Krishna Centre for Development Studies Publication, Bhubaneswar

Samal KC, Meher S, Panigrahi N (2003) Beyond relief disaster mitigation, livelihood rehabilitation and the post-cyclone recovery in Orissa: village level studies in three most cyclone affected districts in Orissa. Nabo Krishna Centre for Development Studies Publication, Bhubaneswar

Sen AK (1999) Health in development. Bull World Health Organ 77(8):619–623. http://cdrwww.who.int/bulletin/archives/77(8)619.pdf. Accessed 02 June 2017

Sen AK (2003) Forward. In: Farmer P (ed) Pathologies of power: health, human rights, and the new war on the poor. University of California Press, California
Sen AK (2009) The idea of justice. Penguin Books, London
Shiva V, Emani A (2000) Climate change, deforestation and the Orissa super cyclone: ecological costs of globalisation. Research Foundation for Science, Technology and Ecology, New Delhi
Sivakumar MVK (2005) Impacts of natural disasters in agriculture, rangeland and forestry: an overview. In: Sivakumar MVK, Motha RP, Das HP (eds) Natural disasters and extreme events in agriculture: impacts and mitigation. Springer, New York, pp 1–22
Srivastava SK (2009) Making a technological choice for disaster management and poverty alleviation. Disasters 33(1):58–81. doi:10.1111/j.1467-7717.2008.01062.x
UN (2005) Hyogo framework for action 2005–2015: building the resilience of nations and communities to disasters. In: World conference on disaster reduction, Kobe, Hyogo, Japan, 18–22 Jan 2005
UN (2015) Sendai framework for disaster risk reduction 2015–2030. UNISDR, Geneva. http://www.unisdr.org/we/coordinate/sendai-framework. Accessed 24 June 2016
UNISDR (2015a) Making development sustainable: the future of disaster risk management: global assessment report on disaster risk reduction 2015. UNISDR, Geneva
UNISDR (2015b) Japan provides early warning example. In: UN world conference in disaster risk reduction 2015 Sendai Japan. Third UN world conference on disaster risk reduction. www.wcdrr.org. Accessed 16 Mar 2015
Visvanathan N, Duggan L, Nisonoff L, Wiegersma N (eds) (1997) The women, gender and development reader. Zed Books, London
Vollmer H (2013) The sociology of disruption, disaster and social change: punctuated cooperation. Cambridge University Press, Cambridge
Wahlström M (2015) From agreement to action: what next for the new global framework on disasters?. The Policy Lab Meeting at the Royal Society, London, 24 June
Whiteford L, Rylko-Bauer B, Farmer P (2009) Epilogue: global health in times of violence—finding hope. In: Rylko-Bauer B, Whiteford L, Farmer P (eds) Global health in times of violence. School for Advanced Research Press, Santa Fe, pp 223–232
WHO (2010) Health sector treats almost 6 million in flood-affected Pakistan, but great challenges persist. WHO. http://www.who.int/mediacentre/news/releases/2010/pakistan_floods_20100923/en/. Accessed 21 Jan 2014
WHO (2013) WHO reference group on global health statistics report of the 1st meeting. WHO. http://www.who.int/healthinfo/statistics/ReferenceGroupGlobalHealthStatistics_MeetingReport1.pdf?ua=1. Accessed 21 Jan 2014
WHO (2016) Infographics on health and climate change. World Health Organization. http://www.who.int/globalchange/climate/infographics/en. Accessed 14 July 2016

Chapter 2
Systems Failure in Disasters

Abstract This chapter explains why deaths in disasters occur by taking two overarching perspectives: risk and vulnerability. Risk or the traditional perspective gives the advantage of understanding the dynamics of geohazards and their effect on humans. Vulnerability perspective on the other hand, helps in explaining why some groups of people are more vulnerable to disasters than others due to their class, gender, age, and race identities. This chapter also adds on an additional perspective to explain deaths in disasters. This is a complex perspective. In this perspective, deaths occur due to the vulnerabilities that exist in the seams of disaster management system. This system is a conglomeration of different professional groupings and actors designed for specific tasks and goals. It is also a system that is highly reliant on technology. As such, loose coordination and communication between actors can lead disaster management system to fail. To showcase, how the disaster management system can fail to save lives, an analytical tool for systems failure is presented with its three inter-connected components: coordination, communication and world views.

Keywords Complex perspective · Coordination failure · Communication failure · Disaster climate · Risk · Vulnerability · Socio-technical system · Systems failure · World views · Wicked problem

The central question of this book is: why do avoidable deaths occur in disasters? Research that scrutinises the causes for human deaths in disasters are rather limited, but they indicate, broadly speaking, two perspectives: traditional (or risk) and vulnerability (Kapur 2010). Both perspectives are analytically distinct but in practice they are related. This chapter also provides a complex perspective to explain avoidable deaths in disasters.

N.S. Ray-Bennett *Avoidable Deaths*, SpringerBriefs in Environmental Science,
https://doi.org/10.1007/978-3-319-66951-9_2

2.1 Traditional Perspective

The traditional perspective is the mainstream or dominant perspective. According to this perspective, natural hazards[1] originate from natural systems and they can cause harm and loss. One way of mitigating the effect of nature is through technology or a 'technical fix' (Ariyabandu and Wickramasinghe 2003; Bryant 2005; Gilbert 1998; Ray-Bennett 2009a; Wisner et al. 2004). This line of thinking was dominant in the UN's General Assembly Resolution 44/236, adopted on 22 December, 1989. Four out of five of its goals underlined the importance of the dissemination of technical information and the transfer of scientific and engineering knowledge for the mitigation of disasters in developing countries (Bankoff 2001; de Senarclens 1997). As a result, structural mitigation measures, such as building concrete houses, flood levies, ocean wave barriers, cyclone shelters, embankments and dams, attained primary importance over non-structural mitigation measures, such as policies, laws, training, raising public awareness and aid—amongst many (Davis and Gupta 1991; Haque and Zaman 1994; Kaiser et al. 2003; Zaman 1999). This technical perspective has evolved due to the mid-term evaluation of the International Decade for Natural Disaster Reduction (IDNDR) (1990–2000) in 1994 (known as Yokohama Strategy) followed by the Hyogo Framework for Action (2005–2015) and most recently the Sendai Framework for Disaster Risk Reduction (2015–2030) (UN 2005, 2015). Now there is widespread acknowledgement that hazards can include "latent conditions that may represent future threats and can have different origins: natural (geological, hydrometeorological and biological) or induced by human processes (environmental degradation and technological hazards)" (UNISDR 2015a: 3/25).

Despite these changes, governmental organisations often use natural causes or the geophysics of a hazard to explain deaths in disasters. This is noted by Kapur (2010) whilst reviewing the effects of 16 natural hazards[2] in India from 1977 to 2002. Geophysics of a hazard can be understood in three ways: First, higher the intensity of a hazard, the more likely it is to kill people. Intensity is classified as moderate or severe for 11 hazards out of 16. Since the focus of the case study is on cyclones, the intensity of a cyclone is determined by its wind speed (such as moderate or severe). Of the 11 hazards cyclones have killed more people in India. Second, hazards are seasonal and so are human deaths. In India, the month of November is cyclone-prone, May is prone to gale and dust storms, April for hailstorms, June for lightning, and January and February for cold snaps. It was noted that 32% of cyclones occurred in the month of November and 36% of all deaths due to disasters were in this month. Third, the effect of hazard is spatially determined and so are the deaths due to disasters are spatially varied. India is

[1]The Sendai Framework for Action defines hazard as "A potentially damaging physical event, phenomenon [...] that may cause the loss of life or injury, property damage [...]" (UN 2015: 3/25).

[2]The 16 natural hazards are: cloudburst, cold wave, drought, dust storm, earthquake, flash flood, gale, hailstorm, heat wave, lightning, snowfall, squall, thunderstorm (Kapur 2010).

diverse and different regions are exposed to different types of hazards. For instance, the coast of Bay of Bengal is exposed to severe cyclonic storms, whereas the north–west is exposed to droughts. Almost one half of all deaths in the Bay of Bengal were due to cyclones compared to the west coast of India (Kapur 2010). The coast of Bay of Bengal has also experienced highest death toll compared to the other coasts in the world. As mentioned earlier, in the 1970 cyclone more than 500,000 people died. In 1991 cyclone almost 140,000 people died in Bangladesh (Haque et al. 2011).

This perspective provides an excellent insight into the dynamics of geohazards and their effect on humans. As a result, national and international organisations are investing heavily to build the capacity of the experts and practitioners by embracing state-of-the-art technologies, such as space technology and multi-hazard early warning systems in order to promote effective disaster management system to reduce deaths (UNISDR 2015b). However, in the context of this research, this perspective explains little as to why more women die in disasters than men or vice versa.

2.2 Vulnerability Perspective

The vulnerability perspective, on the other hand, aims to explain why some people are more vulnerable to disasters than others. Vulnerability is often used in different ways (Bacon et al. 2017) but in this instance, it is understood as "*the characteristics of a person or group and their situation that influence their capacity to anticipate, cope with, resist and recover from the impact of a natural hazard* (an extreme natural event or process)" (original emphasis by Wisner et al. 2004: 11). According to this perspective, the impacts of natural disasters are not entirely 'natural'; rather they are determined by people's unequal exposure to risks which are a consequence of the socio-economic systems (Cannon 1994; Neumayer and Plümper 2007). The significance of natural hazards as trigger-events is not denied by this perspective, but emphasis is placed on the various ways in which social and economic systems can render people more vulnerable to disasters (Cannon 1994; IPCC 2012; Varley 1994; Winchester 2000; Wisner et al. 2004). In this perspective, the normal daily lives of some vulnerable groups are often difficult to distinguish from disaster conditions. Proponents assert that disasters only act as an interface between an extreme physical environment and a vulnerable group of the population, due to a "combination of hazards, conditions of vulnerability and insufficient capacity or measures to reduce the potential negative consequences of risk" (Sivakumar 2005: 2).

According to this perspective, differences in mortality to natural disasters are explained due to biological vulnerability, social and cultural vulnerabilities (caste, race, gender, and age), economic vulnerability (class) and physical vulnerability (housing structures). These vulnerabilities are not distinct; they often are conjoined and reinforce each other during the time of disasters. This is evident in the discussion below.

Biological and physiological differences between men and women put women at high risk during disasters (Ariyabandu 2009; Neumayer and Plümper 2007; Ariyabandu 2009). Men in general are physically stronger than women and therefore they are likely to withstand the impact of a disaster better than women. For instance, a physically robust man has a better chance to swim or climb up a tree in order to survive against an emerging storm surge. However, biological and physiological differences may also be socially determined (Eklund and Tellier 2012; Neumayer and Plümper 2007).

Social and cultural norms related to role behaviour put women, more than men, at a greater risk when it comes to rescue efforts (Neumayer and Plümper 2007). Dress codes such as *saree* or *burqa* were found inhibiting women's mobility during the 1991 cyclone in Bangladesh. Learning to climb a tree or to swim are socially not permissible in some societies. In Sri Lanka, a study conducted by the Centre on the Epidemiology of Disaster (CRED) noted that only 12–20% of women were able to swim compared to 75–85% of men (Eklund and Tellier 2012).

In the response phase, the lack of lifesaving skills along with the responsibility of looking after young children often put women at a greater risk to disasters (Eklund and Tellier 2012; Neumayer and Plümper 2007). A report by the WHO (2002) noted that although the Indian Ocean Tsunami in 2004 claimed 300,000 lives across 13 nations; 80% of these lives were women and children. The Indian Meteorological Department found that a large proportion of women (34%) and children (45%) were killed in disasters in comparison to men (21%) between 1977 and 2002 (Kapur 2010). Neumayer and Plümper's (2007) research of 141 countries between 1981 and 2002 showed that natural disasters lower the life expectancy of women more than that of men. In other words, natural disasters (and their subsequent impact), on average, kill more women than men and kill women at an earlier age than men. This study also observed:

> The stronger the disaster (as approximated by the number of people killed relative to population size), the stronger this effect on the gender gap in life expectancy. The higher women's socio-economic status, the weaker this effect on the gender gap in life expectancy (Neumayer and Plümper 2007: 1).

Disproportionate numbers of women's deaths were also reported in Mexico City in the 1985 earthquake and the Bhuj earthquake in Gujarat in 2001, and more elderly women's deaths in the Kobe 1995 earthquake (Hewitt 2013; Kapur 2010). In 2005, the earthquake in Pakistan killed three times more women than men (Trust Org 2013). Women accounted for 61% of deaths in Cyclone Nargis in 2008 in Myanmar (Aus Aid 2014) and 67% in Banda Aceh in the Indian Ocean Tsunami in 2004 (GFDRR, quoted in Bradshaw and Fordham 2013). Death rates were almost four times higher amongst women than men in the 1991 Bangladesh cyclone (Aus Aid 2014). Women's mortality rates are also much higher in developing countries in comparison to their developed counterparts (Lass et al. 2011; Neumayer and Plümper 2007).

Gender based division of labour could be equally detrimental to the effect of some hazards. For instance, more women were killed in the Latur Earthquake in

1993 because during the time of the earthquake women and children were at home, whereas men were out working in the field (Kapur 2010). In contrast, more fishermen and male labourers were killed in the district of Jagatsinghpur in Odisha during the Super-Cyclone 1999 (Samal et al. 2003). The coastal livelihoods in Odisha are reliant heavily on fishing and prawn cultivation. Prawn cultivation is a labour-intensive activity and employs largely male labourers. When the Super-Cyclone made its landfall, most men were either fishing in the sea or were involved in rearing of the prawns near to the coast. Consequently, they were directly on the pathway of the Super-Cyclone. This is consistent with the research findings. In this light, social norms and roles provides an additional layer to the biological and physiological reasons for the gendered nature of disaster vulnerability. Women are physically weaker than men but men may be more vulnerable to certain types of disaster risks than women (Neumayer and Plümper 2007).

Social stratification based on class and race can be highly detrimental in putting some lives at risk during disasters (Barnshaw and Trainor 2007). An individual's class and race often determines the choice one can make and the social capitals they can acquire in their everyday lives. Social choices and capitals are critical assets to build individual's agency and social network. Vulnerable groups often lack agency due to structured inequalities and lack of social resources that exist in their everyday lives. In this vein, communities which are better placed with social networks will be in a better position to evacuate and even survive a disaster. In contrast, communities which are not will be more susceptible to the impacts of the disaster. For example, higher numbers of poorer African–Americans were the victims in Hurricane Katrina compared to their white middle class counterparts because they lacked social capital in their everyday lives (Barnshaw and Trainor 2007).

Caste and class is an equally important determinant in the Indian Subcontinent for gender differences in disaster mortality. In fact, the relationship between caste, class and gender in Indian society cannot be understood outside a consideration of their mutual impact (Chakravarti 2003); they are intertwined and interlinked, creating and reinforcing inequality (Sen 2003). According to Sen (2003: 207):

> Class does not act alone in creating and reinforcing inequality, and yet no other source of inequality is fully independent of class. Consider gender. […]. Gender is certainly an additional contributor to societal inequality. [ ]. Similarly turning to caste, even though being lower caste is undoubtedly a separate cause of disparity, its impact is all the greater when the lower-caste families also happen to be very poor.

Caste, class and gender issues were noted by the author (Ray-Bennett 2009a, c) in the Super-Cyclone of 1999 in the village of Tarasahi, Odisha. She found women-headed households from the upper and middle castes were better able to survive the Super-Cyclone than their low caste counterparts. This was because the low caste women lacked social support. Their houses were also not made from concrete and were located in the low-lands which were the first to get flooded. Lack of evacuation shelters further exacerbated the predicament of the low caste women. The low caste respondents survived the impact of the Super-Cyclone by tying onto a coconut tree, squatting for more than 10 h by holding hands tightly with their neighbours under a plastic sheet, whilst others survived by sitting on the veranda of a rich upper caste household and of the public school buildings.

In addition to the social and cultural vulnerabilities, physical vulnerability can also lead to gender differences in mortality. Kapur (2010) argued that greater numbers of women's deaths occurred in the 1993 Latur Earthquake in India due to the nature of house structures. Higher numbers of deaths were recorded in stone and mud houses (86.32%) compared to shacks (0.40%) and brick and mortar (1.15%). Although poor households owned mostly the stone and mud houses and therefore bore the severe brunt of the Earthquake, it is important to note that not all of the poor suffered, because the households living in shacks were the least affected. This indicates that poverty and disaster vulnerability are linked but that they are not the same (Cannon 1994; Jaspars and Shoham 1999; Ray-Bennett 2009a). The correlation of vulnerability and poverty is highly significant, but concomitantly failure to distinguish vulnerability from poverty has severe policy implications because poverty is endemic and defined by professionals in terms of flows of income and consumption (Chambers 1989). Anti-poverty programmes tend to concentrate on raising incomes or consumption and progress is measured according to these flows, which are then often taken as indicators of other dimensions of deprivation, including vulnerability (Chambers 1989). Poverty is largely a consequence of class and social position and in itself provides an inadequate explanation of the differential impact of hazards (Cannon 1994; Ray-Bennett 2009a). Men and women's deaths in disasters when attributed to their class and social vulnerabilities cannot be accepted uncritically—other vulnerabilities play a part too.

2.3 Complex Perspective

It can be argued that human deaths can occur not only due to the vulnerabilities that exist in the natural and social systems, as discussed above, but also due to vulnerabilities that exist in human built organisations[3] (Perrow 1999; Weick 1990), such as the disaster management system. A disaster management system is a conglomeration of different professional groupings and actors designed for specific tasks and goals. It is a system that is highly reliant on technology. Also, actors[4] within this system adopt different frames of reference. As such, weak forms of organisation between actors could potentially lead to systems failure causing deaths

[3]"Organisations are social designs directed at practice" (Wenger 1998: 241). In other words, organisations are combination of institutions (social design) and constellation of practices by different actors which gives life to the organisation (Wenger 1998).

[4]Actors are "the agents who carry out or cause to be carried out the main activities of the system, especially its main transformation" (Checkland 1981: 224). There are also victims or beneficiaries in these soft systems. In this study, they are the vulnerable groups of men, women and children who are indirectly involved in this research.

Table 2.1 Typology of organisations and actors

Organisations	Actors
Primary	Category 1 responders and affected community
Secondary	Category 2 responders
Tertiary	Global responders

in disasters. This is discussed through the analytical tool of systems failure later in this chapter.

Vulnerabilities at the seam of a disaster management system can emerge due to the internal vulnerabilities of actors, which can manifest in the form of inaction, human errors, mismanagement, lack of coordination, hierarchy, communication—to mention a few. This is because disaster management involves multiple agencies and actors across the public, private and voluntary sectors at local, regional, national and global levels. These organisations are diverse, hierarchical and inter-dependent. A typology of these organisations and actors is provided in Table 2.1. In the case of an external threat such as natural disaster or political disaster, these internal vulnerabilities come to the fore causing the systems to fail.

Primary organisations are the first responders. They are also known as Category 1 responders by the Civil Contingency Act in the UK (Walker and Broderick 2006). They have specified responsibilities—"risk management, emergency planning, business continuity of the responder itself, and warning and informing the public" (Walker and Broderick 2006: 81). Although much of the 'communication, co-operation and information sharing' falls within the remit of the Category 2 responders or secondary organisations—responders are also expected to share information with other local responders to enhance co-ordination and co-operation. Communities who are directly affected may also be seen as first responders (Kolen and Helsloot 2012; Quarentelli 1977).

Secondary Organisations are decision making organizations at national and state levels. They undertake a myriad of roles and responsibilities which can be categorised as auxiliary, alleviating and collateral. Auxiliary organisations have some form of interactive contact with the primary organisation prior (Toft and Reynolds 2005) to a disaster climate. They make decisions, raise disaster funds, coordinate relief activities and share information with primary organisations to enhance response. Examples include the Ministry of Rural Development, the Ministry of Home Affairs and the Disaster Mitigation Authority in India who work closely with the state level disaster management authorities (Srivastava 2009). Alleviating organisations (Toft and Reynolds 2005) complement the primary and auxiliary responders, businesses and the public by generating valuable early warnings for decision making and by planning for response. Examples include the Indian Meteorology Department, the Meteorology Office in the UK and the Pacific Tsunami Warning Centre in Hawaii, to name a few.

Collateral organisations (Lalonde 2011: 450), are "new form of [secondary] organisations that do not replace the usual operational structure but co-exist and contribute to the problems that are non-routine". For example, the Odisha State

Disaster Management Authority, Gujarat State Disaster Management Authority and the Disaster Mitigation Authority under the auspices of the Prime Minister in New Delhi—all exist solely to address disaster management and mitigation issues. Other examples include the Environment Agency in the UK and the Federal Emergency Management Authority in the United States.

Tertiary organisations are multi-lateral or global organisations formed between "three or more nations to work on issues that relate to all of the countries in the organization" (Global Energy Network Institute (GENI) 2014). Examples include the UN, and the International Strategy for Disaster Risk Reduction (ISDR)—a dedicated secretariat established by the UN General Assembly in December 1999 for disaster risk reduction. Global financial institutions, such as the World Bank, are also tertiary organisations.

2.4 A 'Wicked' Problem—Deaths in Disasters

According to the complex perspective, it is argued that actors and organisations involved in disaster management belong to linear systems. They are linear because these systems are spatially segregated (Perrow 1999). They have dedicated connections, and have extensive understanding of the nature of risk (such as cyclones/flooding) (Perrow 1999). They are also highly complex in their structures and processes of work. At the outset, they may look like separate entities but in reality, they are highly inter-dependent to each other (Weick 1990). In this context, deaths in disaster are a complex problem. Their complexity arises because the decision to save lives during disasters sits across different governmental departments and institutions (Grint 2008; Rittel and Webber 1973). Some institutional arrangement for disaster management and decision making in Odisha was introduced in Chap. 1—a compilation of this is presented in Table 2.2. As such, deaths in disasters are understood as 'wicked problems' because they are unique with no prior precedents (Rittel and Webber 1973). They involve (or are perceived to involve) poorly understood problems, and require examination at the seams of

Table 2.2 Disaster management institutional arrangements in Odisha (Compiled by author)

• Revenue and Disaster Management Department (Secondary/Category 2)
• Special Relief Organisation (Secondary/Category 2)
• Odisha State Disaster Management Authority (Secondary/Category 2)
• State Disaster Management Authority (Secondary/Category 2)
• The State Executive Committee; Natural Calamity Committee (Secondary/Category 2)
• District Disaster Management Authority (Primary/Category 1)
– Block Disaster Management Committees
– Gram Panchayat Disaster Management Committees
– Village Level Task Force Committees
• Indian Meteorology Department (Secondary/Category 2)
• UNDP (Tertiary/Global)
• International NGOs, International Federations (Tertiary/Global)

multiple disciplines (Midgley 2014), including the sociology of risk and crisis, geography, management, development, violence/conflict, disaster risk reduction, justice and gender (some of which are discussed in this and the previous chapter) to understand the high degree of complexity in order to reduce deaths.

Wicked or complex problems have their roots in systems thinking[5], particularly a soft systems thinking (Bunge 1977; Midgley 2014; Rittel and Webber 1973). "Soft systems typically have properties that are difficult to quantify and measure e.g. viewpoints, conflicts, vested interests and other qualitative aspects" (Waring 1989: 11). In soft systems, human activity is the key. This view has its roots in Weber's interpretive social science (phenomenology and hermeneutics), Dilthey's *Weltanschauung* (W) (world view), Husserl's 'idea of phenomenology' and Vicker's appreciative system.[6] As such, wicked problems can be explained in numerous ways, but the choice of explanation will determine the nature of the problem's resolution (Rittel and Webber 1973). Wicked problems involve a level of uncertainty and ambiguity which requires political collaboration and 'adaptive leadership' as opposed to scientific processes only (Grint 2008). Wicked problems have no right or wrong solutions (Grint 2008; Rittel and Webber 1973). Also, "problems are not characterised by linear causal chains, rather they are circular, emergent and continuously changing in a messy, ill-structured, real-world" (Checkland 1985: 298). Accordingly, in such systems, human problems tend to be persistent ones. They are influenced by humans' world views, and different points of view, where the only way to resolve issues is to seek 'accommodation' rather than 'solutions' (Checkland 1985). This is discussed at length in the ensuing sections through the analytical tool of 'systems failure'.

---

[5]Systems thinking can be understood in terms of 'three tightly interrelated discourses: general systems theory (GST), cybernetics and complexity (Midgley 2003: xxii). These theories emerged in the mid and late twentieth century. All the three discourses are highly interlinked and favour mathematics and modelling as their systems language (Checkland 1981; Midgley 2003). As a result, organisational and management theorists have applied only a 'partial systems approach' because of the nuances and dynamics that underpin a human system (for detailed discussion see Kast and Ronsezweig 1972; Midgley 2003; Mingers 1980).

As an alternative to the aforementioned theories, Checkland (1981, 1985) conceives systems as one of two types: hard and soft. A hard system is a 'goal seeking' engineered system. Checkland (1981, 1985) views general systems theory and cybernetics as hard systems. Hard systems have quantifiable and measurable attributes (Checkland 1981, 1985; Waring 1989). Soft systems, on the other hand, are not goal seeking engineered system, rather they deal with 'soft' problems in social systems where goals are often obscured, ambiguous or non-existent (Checkland 1981, 1985). Soft systems deal with the complex problems of an ill-structured and poorly understood real world.

[6]Max Weber's interpretive social science became popular with the emergence of phenomenological sociology in the 1960s. Phenomenology like positivism is also a philosophy, which was made popular by Edmund Husserl in the early part of twentieth Century (McNeill and Chapman 1985). For Husserl, 'the basic reality lies in our thinking' and 'the everyday we take as given is in fact constructed through human activity' (Checkland 1981: 274). Alfred Schutz also applied Husserl's idea of phenomenology to the study of social life or *Lebenswelt* (lived-in-reality) (McNeill and Chapman 1985). According to Checkland (1981) Vicker's appreciative system is based upon Schutz's *Lebenswelt*.

2.5 Systems Failure

Studies on systems failure stem from cognitive psychology, organisational culture, risk management and systems engineering theories. These studies examine the internal vulnerabilities of actors and organisations as a contributing factor for systems failure broadly under three approaches (Dekker 2006); namely the normal accident approach, the human error approach, and the systemic approach. A short review of these approaches is provided below in order to make the case for 'systems failure'.

The normal accident approach: Organisational theorist Perrow (1999) challenged the traditional causal factors of 'operator error' in system accidents. He did this in the context of high-risk technologies or 'risky enterprises[7]'. High risk systems are especially complex because they work in human-technology interface. Perrow offered two special characteristics of high risk systems: 'interactive complexity' and 'tight coupling'. In the conditions of interactive complexity:

> when x failed, y would also be out of order and the two failures would interact so as to both start a fire and silence the fire alarm. Furthermore, no one can figure out the interaction at the time, and thus know what to do. [...] This interacting tendency is a characteristic of a system, not of a part or an operator (Perrow 1999: 4).

'Tight coupling', on the other hand, exists when:

> processes happen very fast and can't be turned off, the failed parts cannot be isolated from other parts, or there is no other way to keep the production going safely (Perrow 1999: 4).

System accidents include multiple factors or a chain of events (concatenation) that may not initially seem to be the 'root causes' of the failure. These chains of events are connected either through linear or complex interactions which could be tightly or loosely coupled (Perrow 1999; Weick 1990). This is also termed 'the sequence-of-events' model (Dekker 2003, 2006).

The human error approach: Reason (1990), a cognitive psychologist, also examined system accidents in risky enterprises, but focussed more on 'human error[8]' than technology. Reason offered two categories of human errors: active and latent. Active errors are those "whose effects are felt almost immediately", whereas latent errors are those "whose adverse consequences may lie dormant within the system for a long time, only becoming evident when they combine with other factors to breach the systems" (Reason 1990: 173). This approach is also known as an 'epidemiological' one in which latent errors act as resident pathogens (Dekker 2006).

---

[7]Risky enterprises include nuclear power plants, chemical plants, the energy sector and the mass transportation sector (including road, rail, sea and air).

[8]According to Reason (1990: 17), the term error can only be "meaningfully applied to planned actions that fail to achieve their desired consequences without intervention of some chance or unforeseeable agency".

In 1997, Reason proposed 'organisational culture' as another important tenet of systems failure. Organisational culture, in general, is a problematic concept. This is because there is no single culture but rather subcultures often formed around the interests of particular professional groups. According to Reason, organisational culture is "shared values (what is important) and beliefs (how things work) that interact with an organisation's structures and control systems to produce behavioural norms (the way we do things around here)" (Uttal is quoted in Reason 1997). To avoid organisational accidents, Reason advocated a 'safety culture', a culture which is flexible, just and promotes reporting. "Together they interact to create an *informed culture*" (Reason 1997: 196, original emphasis).

The concept of 'safety culture' was further extended by risk specialists Toft and Reynolds (2005). They viewed organisations as socio-technical systems, and socio-technical failures as a combination of human and technical failures. They understood safety culture as key to reducing socio-technical disasters but noted that it is often not prioritised by the organisation. This is partly because safety culture can be obstructed by politics and a 'culture of blame'. The 'culture of blame' is rife in businesses and organisations. Proponents of organisational and risk management studies demonstrated this by understanding organisations as highly political and politicised spaces (Blockley 1996; Douglas 1966; Grey 2009; Horlick-Jones 1996; Weir 1996).

Systems accidents often involve human loss. Instances include the Challenger space shuttle (seven deaths), Zeebrugge ferry sinking (193 deaths) (BBC 2014), King's Cross underground station fire (31 deaths) (Fennell 1988), and Bhopal gas tragedies (3,800 deaths) (Broughton 2005)—to mention a few. This engages human sympathy as well as socio-political consequences (Weir 1996). As a result, these involve a perceived necessity to apportion blame. In this light, organisations instantaneously pursue after-the-event explanations in terms of operator error (Weir 1996) because it is a cheap and easy approach in comparison to inviting systemic investigation in a resource constrained environment, which might encounter time, money and political constraints (Dekker 2006). Also, admission of systems failure opens up the organisation to the charge of corporate manslaughter. In this way, 'blaming operators and protecting the interests of designers' also becomes a cause of future system failures (Horlick-Jones 1996). One direct consequence of the blame game is that it hinders organisational learning and increases the likelihood that human errors and mistakes will occur in future. To counteract this, Horlick-Jones (1996) proposes a 'no blame culture', much like Reason's 'safety culture'.

The systemic approach: Dekker (2003, 2006), a cognitive systems engineer, argues that so-called human error flags an opportunity to investigate systemic problems. According to Dekker, systems are not safe, but people make them safer through their practices, experiences and reflection. Accidents are not seen as abnormal in this approach, rather they are "structural by-products of a system's normal functioning" (Dekker 2006: 17). Therefore, system failures deserve systemic investigation in order to draw relevant lessons and become more resilient.

All these studies shed new light on this research. In the context of this research it is argued, systems can also fail due to the *problems of coordination, communication* (Weick 1990) and *conflicting world views*. Unlike high risk systems discussed above, the components of systems failure are built in the context of linear systems, such as disaster management to understand deaths in disasters. These components are discussed after presenting three interconnected assumptions for systems failure which are: interdependencies of systems, disaster management as a socio-technical system, and a disaster climate is an opportune time to observe the manifestation of these two.

Interdependencies of systems: According to systems thinking the domain of disaster risk management can be understood as natural,[9] human[10] and technological systems[11] (Smith and Petley 2009; Tanaka 2015). In reality, these systems are interconnected. Human activities such as rapid population growth, especially in disaster-prone areas and continued mass urbanisation, much of which is unplanned and unsafe (Department for Foreign and International Development (DFID), 2013) both exacerbate the effects of natural weather phenomena by contributing to global warming and climate change (Gillies 2014; Intergovernmental Panel for Climate Change (IPCC) 2007, 2012; Smith and Petley 2009; UNDP 2007; UN 2015). Human systems are then an integral part of ecological or natural systems (Vickers 1983).

The state-of-the-art technologies for weather forecasting, engineering structures to contain hydro-meteorological risks and international strategies and national policies/programmes for disaster risk management are the combinations of hard and soft systems developed by the human minds to minimise disaster risks. In the 'hierarchy of systems complexity' (Checkland 1981; Jenkins 1969), Human Activity Systems (H.A.S) or human systems are at the apex (Checkland 1981). Therefore, natural, social and technological, "are human systems since they are distinguished by human minds and judged to be acceptable by their correspondence with human standards" (Vickers 1983: 210).

---

[9]According to Checkland (1981: 110), natural systems are: "Physical systems [which] apparently make up the universe. These range from the subatomic systems of atomic nuclei as described by physics and the living systems observed on earth to galactic systems at the other extreme. All these are natural systems, systems whose origin is in the origin of the universe […]".

[10]Human systems are part of social systems. According to Vickers's (1983: 216), human systems are relationship maintaining systems that come "into being by their actions and their experiences". These systems are highly political, another man-made element (Tanaka 2015; Vickers 1983). They are also by far the most complex systems (Vickers 1983).

[11]Engineered systems, on the other hand are technological systems. Examples include forecasting and early warning systems and structural mitigation measures. Technological systems are constructed by humans through science (Waring 1989), but technology cannot exist on its own. Both human activity and engineered systems exist in conjunction with each other. Emery (1993) explains this phenomenon as socio-technical systems. Socio-technical systems comprise the 'technological', which is the work and procedural activities that are undertaken, and the 'social', which relates to the "social structure consisting of the occupational roles that have been institutionalized in its use" (Emery 1993: 296).

Socio-technical disaster management system: Traditionally disaster management is understood as a continuum of interlinked activity; it is not a series of events which start with each disaster occurrence, rather it involves the composition of post disaster review, results of exercises of simulations, prevention, mitigation, preparedness, disaster impact, response, recovery (restoration, rehabilitation, reconstruction), and development (Carter 1999). "Disaster management processes are enacted once the immediacy of the disaster event has become evident and resources and capacities are put in place with which to respond prior to and following impact. These include the activation of early warning systems, contingency planning, emergency response (immediate post-impact support to satisfy critical human needs under conditions of severe stress), and, eventually, recovery" (IPCC 2012: 35). Due to the Hyogo Framework for Action and currently the Sendai Framework, the concept of disaster management has changed—from traditional disaster management (top down, technical, exclusion of communities at risk) to an alternative disaster risk management (inclusive of all aspects of risk reduction, disaster preparedness, immediate relief, rehabilitation and long term construction) and most recently disaster risk governance (UN 2005; 2015). However, there still exists ideological gaps in conceiving disaster management as a socio-technical disaster management system, one in which humans and technology interface. See Fig. 2.1 on Socio-Technical Disaster Management System.

In socio-technical disaster management system actors, organisations and technology are interdependent but these interdependencies are less understood in linear systems because they are spatially separated (Weick 1990). This is illustrated whilst developing the components of systems failure, which are coordination, communication and world views related to early warnings and their dissemination. When the system's interdependencies are not understood fully by the relevant actors, the system fails (Weick 1990) in a 'disaster climate' with a devastating consequence of human losses. This leads into the third assumption of a 'systems failure'.

Disaster climate: Of all the climatic hazards,[12] tropical storms/cyclones and coastal flooding[13] are notorious for taking more human lives (Dilley et al. 2005; Haque et al. 2011; Kapur 2010; Nathan 2009). Most of the causalities in these hazards occur at the last phase of disaster preparedness and early phase of response (Prizza 2007). Any deaths[14] caused by the direct impact of these disasters are called

[12]Climatic hazards include flood, cyclone, drought and localised storms (Burton et al. 1993).

[13]Tropical storms are "heavy rains followed by tropical storms. They are also one of the most common causes of floods. Storms form over the warm waters of the tropics. These storms are full of moisture. Under the right conditions these giant storms move towards the land, causing a heavy rainfall. This heavy precipitation causes the streams and rivers to overflow leading to inland floods [...]. Coastal flooding usually occurs as a result of severe storms, either tropical or winter. Ocean waves intensify on the open ocean, and these storms make surface water much choppier and fierce than normal. Raging winds can create huge waves that crash on unprotected beaches" (Modh 2010: 6).

[14]Death is defined as 'number of people who lost their life because of the event happened' (Integrated Research on Disaster Risk 2015: 9). 'The number of deaths is the sum of direct and the

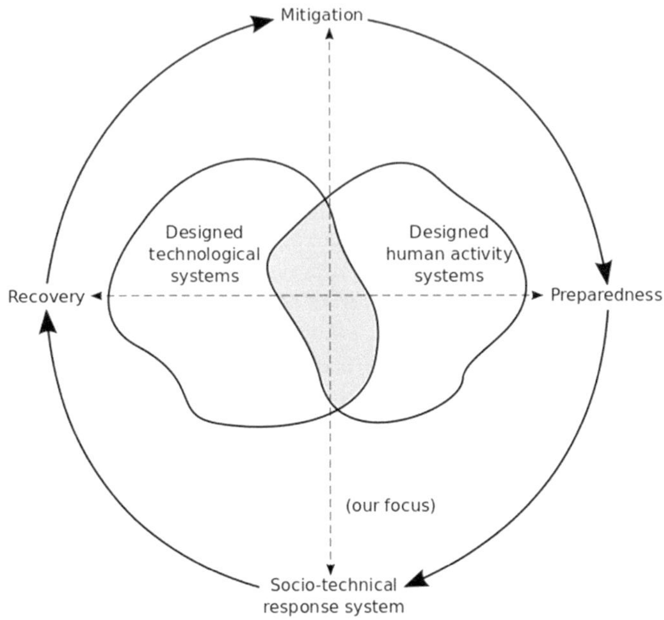

Fig. 2.1 Socio-technical disaster management system. (Produced by author)

'primary deaths' and as evident in the case of Odisha, these deaths are far more widely reported than 'secondary deaths' (Integrated Research on Disaster Risk 2015; Ray-Bennett 2010). Secondary deaths are a consequence of injury or morbidity arising from the impact of disasters. These are the most contentious deaths (Ray-Bennett 2010).

Although disaster management is a continuum of interlinked activity, disaster managers and practitioners understand this involving four phases of mitigation, preparedness, response and recovery[15] (see Fig. 2.1) (Comfort et al. 2004; Miles 2012). Since most of the casualties occur at the last phase of preparedness and early

(Footnote 14 continued)

indirect deaths. The number of delayed indirect deaths is generally excluded. The number of deaths does not include missing persons' (Integrated Research on Disaster Risk 2015: 9).

[15]Mitigation is understood as "The lessening or limitation of the adverse impacts of hazards and related disasters" (UNISDR 2009a: 19); preparedness is understood as "The knowledge and capacities developed by governments, professional response and recovery organizations, communities and individuals to effectively anticipate, respond to, and recover from, the impacts of likely, imminent or current hazard events or conditions" (UNISDR 2009a: 21); recovery as "The restoration, and improvement where appropriate, of facilities, livelihoods and living conditions of disaster-affected communities, including efforts to reduce disaster risk factors" (UNISDR 2009a: 23); and response [our emphasis] as "The provision of emergency services and public assistance during or immediately after a disaster in order to save lives, reduce health impacts, ensure public safety and meet the basic subsistence needs of the people affected" (UNISDR 2009a: 24).

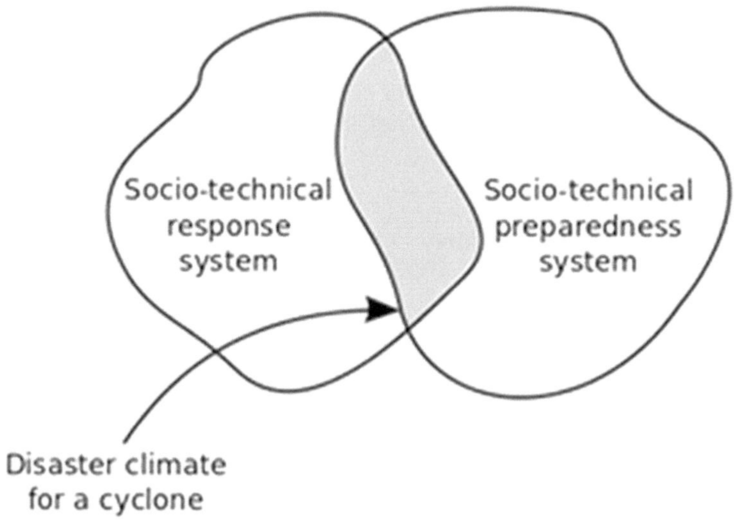

Fig. 2.2 Disaster climate for a cyclone. (Produced by author)

phase of response for the chosen hazards, preparedness and response phases are conjoined. The intersection of these phases make that particular moment in time the most dynamic phase. This phase involves *uncertainties* about the nature of the hazard, as well as conditions conducive to poor decision making. Human actions, decisions for evacuation and the like, communications (strengths and limitations of organisational vulnerability and resilience) and the early warning systems all play a crucial role in avoiding deaths in this dynamic phase. Leadership, timing and readiness are other important factors in this phase. The conditions in this dynamic phase are referred as the 'disaster climate'. See Fig. 2.2 on disaster climate.

The disaster climate occurs as the hazard appears with the potential to affect not just the people who are at risk, but also actors, organisations, infrastructure, ecosystems and communication. The extent of these impacts will also vary both locally and spatially. In such a climate, the first responders begin to operate under an incident management system (Comfort et al. 2004; GoO 2014a, b; Miles 2012). This phase is typically 48 h prior to the landfall of a cyclone (Kolen and Helsloot 2012). During the landfall of a cyclone, response is usually very difficult (Kalsi 2003; Kolen and Helsloot 2012). In such a context, the chances of safety failure are also high. This is illustrated through the mode of coordination and communication failures later in this chapter.

Disaster management is then a highly complex socio-technical system. The complexity of this system is played out through the network space (Latour 2005) of actors and organisations in a disaster climate. This network space includes disaster management organisations, actors, doctrines, policies, culture, and world views at the interface with technology.

2.6 Components of Systems Failures

Having discussed a few assumptions of systems failure, this section discusses the three inter-connected components of systems failure.

2.6.1 *Coordination Failure*

Disaster management organisations, such as primary, secondary and tertiary, undertake a myriad of activities; it is not possible to track them all. For the purpose of this research, the focus is on coordination problems in a disaster climate.[16] Disaster climate, offers a 'window of opportunity' to identify uncoordinated activities and greater opportunities to learn lessons. Coordination studies have focussed on coordinating human personnel, the division of labour by function (Prizza 2007) and stakeholder partnerships (Chatterjee et al. 2010). In the context of this research, coordination of a flow of 'core information' in order to save lives during disasters, is focussed upon.

Core information is early warning information and constitutes the near-real-time information about the impact of an event. This information is largely generated by meteorologists and meteorology offices using early warning systems.[17] The interpretation of this core information involves a complex interaction of humans and technology (Alexander 1993). Meteorologists use many different tools to generate early warnings for different types of hazards. For climatic hazards (such as cyclone and flooding), geographic information systems (Boehnert 2009) and remote sensing are widely used (Herrmann 2009). These tools are a product of the information age[18] (Alberts and Hayes 2003) and are able to predict the formation of cyclonic depressions as well as provide almost real time tracking of the landfall of a cyclone inland (Kalsi 2003; Srivastava 2009). Core information is valuable to avoid unnecessary deaths and also increase the efficacy of the response system (Comfort et al. 2004).

In this context, problems of coordination can occur when there is a lack of core information. For example, prior to the Indian Ocean Tsunami in 2004, there was no early warning system in place to monitor the Indian Ocean water surface. As such,

---

[16]The author is inspired by Hayek's idea of inter-temporal discoordination (Garrison and Kirzner 1987). Whilst exploring the coordination problem, he emphasised looking at inter-temporal discoordination of economic activities in crisis. This was because it is not possible to track the myriad economic activities that individuals and organisations undertake in everyday life.

[17]UNISDR (2009a) defines early warning systems (EWSs) as: "The set of capacities needed to generate and disseminate timely and meaningful warning information to enable individuals, communities and organizations threatened by a hazard to prepare and to act appropriately and in sufficient time to reduce the possibility of harm or loss".

[18]According to Alberts and Hayes (2003), information age is the successor to industrial and post-industrial ages.

there was no core information for the Category 1 and 2 responders to rely on. The consequence of this was a death toll of more than 225,000 people (Channel 4 2015; Romo-Murphy et al. 2011). In the words of Kofi Annan in 2006:

> If an early warning system had been in place when the tsunamis of 26 December 2004 struck the Indian Ocean region, many thousands of lives could have been saved. That catastrophe was a wake-up call for governments and many others about the role early warning can play in avoiding and reducing the human and physical impacts of natural hazards (UNISDR 2009b: 62).

The lack of core information can blind a response system with the potential to hamper efforts to save lives (Comfort et al. 2004).

The problem of coordination is also closely connected to decision making. Decision making is both a cognitive and social process at the interface of technology in a disaster climate. It is a social process because the knowledge of core information can be a political tool for some interest groups (Alberts and Hayes 2003; Herrmann 2009). The early warning systems are often nestled in bureaucratic units; "embedded within larger bureaucratic units which are themselves embedded within even larger bureaucratic units" (Herrmann 2009: 23). Lifesaving information can be power and information received by one special interest group such as a government ministry for instance since receiving such information ahead of others can be politically very rewarding (Herrmann 2009). For example, the US AIDs Famine Early Warning Systems Network (FEWS Net) warned of an impending crisis in the Horn of Africa as early as August 2010. This was about seven months prior to the drought crisis in 2011, yet the decisions and actions were still delayed for political reasons (Aalst et al. 2013; Global Emergency Group undated). In this context, the core information was available but the decision to develop the response system was hampered by bureaucracy and a lack of political will (Herrmann 2009).

The decision to coordinate 'core information' can also be hindered by a culture of complacency (Carnegie Foundation expert panel members, Acton and Hibbs 2012), poor administrative structures (Chatterjee et al. 2010; UNISDR 2006) and a lack of adaptive leadership amidst the convoluted roles and many actors present in a DRR climate (multi-stakeholder and multi-sectoral approach) (Aalst et al. 2013; Global Emergency Group undated; Lalonde 2011; Prizza 2007).

Second, looking at decision making from a cognitive perspective, coordination problems can arise under conditions of 'uncertainty' (Kolen and Hesloot 2012). Uncertainties and ambiguities are inevitable when dealing with complex problems (Grint 2008). For example, although early warning systems have greatly advanced, it is still not possible to provide exact forecasts of the landfall of a tropical cyclone or exact predictions of flooding.

> Although the development from a tropical depression into a tropical cyclone usually occurs in 12 to 24 hours, 15 percent required more than 48 hours, and other are reported to undergo formation in less than 12 hours [...] [based on temporal disturbances] (Kalsi 2003: 49).

In such uncertainties, decision makers and practitioners could be left without "a reliable warning of a potential threat with enough lead-time to take appropriate evasive action" (Herrmann 2009: 14). In such a dynamic situation, decision makers

(Category 1 and 2) have to cope and react with limited information and uncertainties about potentially fatal conditions.

Some key uncertainties may include assessing potential impacts once the disaster has been predicted, assessing who should be warned, about what and how (Alexander 1993), assessing whether to order a mass evacuation, where to, how, when and using what mode of transport (Kolen and Hesloot 2012), assessing the health, safety and security of evacuees in emergency shelters (Rashid and Michand 2000; Ray-Bennett 2009b) and assessing when to cease warnings (Kalsi 2003; Kolen and Hesloot 2012). In the recovery phase, decisions are made about whether to review, test and modify the system's functionality (both hardware and personnel) in order to address lessons identified (Alexander 1993).

2.6.2 Communication Failure

It is also argued that systems can fail due to deficient communication. According to Anderson and Goolishian (1988), human systems are communicative systems. This communicative sub-system is socio-cultural and is organised 'according to role and structure', that exist in the 'domain of meaning' (Anderson and Goolishian 1988: 187). Human systems are "language-generating, meaning-generating systems engaged in an activity" (Anderson and Goolishian 1988: 188). One can view the early warning systems as language-generating and meaning-generating systems whose objectives include "detection and warning, communication and response" (Kalsi 2003: 68). The relevant actors in a socio-technical disaster management system generate core information at the interface with technology. This technical information is then interpreted in order to communicate across actors and organisations with the aim of supporting a response system. The relevant actors are then generating, as well as communicating, core information between and across actors in order to minimise deaths in the at-risk communities. Through this process, the first responders make sense of this core information and develop an appropriate disaster response (Arnoldi 2001). As such, core information and the disaster response are tightly coupled. Loose coupling (either due to the non-availability or miss-interpretation of the core information) will lead to an ineffective response (Hanai 2014).

Luhmann (1993, 1999) also emphasised the importance of communication in social systems. To him, communication is not just 'a direct transmission of meaning or information between persons', rather it is a closed system in relation to which the persons communicating belong to the environment. According to Luhmann (1999), society and its sub-systems consist of communication only and there is nothing 'social' outside this. The domain of sub-systems also consists of communication and the boundary of these sub-systems 'is the boundary of its observation' (Arnoldi 2001). In this light, the language-generation and meaning-generation of the early warning systems happen within the boundary of disaster risk management.

Communication is then *everything* in a disaster management system. Without communication between the relevant actors, actors to technology and likewise, the function of this sub-system to reduce societal loss and damage from disasters will collapse. This is explained once again through the flow of the 'core information' generated by the early warning systems. The flow of core information can fail for a number of reasons. The most relevant in the context of this research are physical disruption of the early warning systems and the communicating devices, hierarchy and the (in)accuracy of the information.

Physical Disruption: In the information age, disaster management organisations, including the meteorological offices, use a number of information and communication tools or devices and social media to communicate, exchange and share core information amongst themselves and with the at-risk population (Alberts and Hayes 2003; Moore and Verity 2014; Srivastava 2009). Some of these include telephone, e-mail, satellite phone, mobile, TV, radio, newspaper, paging devices, twitter and the Internet. These communication devices are central to improving the capacity of the first level responders (Comfort et al. 2004), as well as that of the at-risk population (GIZ 2012; Harun-Al-Rashid 1997; Moore and Verity 2014; UNISDR 2006). However, a random failure of the communications networks or of power supply caused by a disaster could significantly damage the flow of information between organisations and with the people at-risk (Srivastava 2009).

Disaster management systems are underfunded worldwide. The first level responders often lack in human resources, budgets and communication devices which are essential for developing effective response systems. According to Aid Data, of the total US$4.5 trillion for developmental assistance, only 2% was spent on disaster related activities between 1980 and 2009 (Aalst et al. 2013). Of that 2%, only 3.6% was spent on disaster prevention and preparedness. Even when core information can reach the first responders, budgetary constraints mean they may lack the resources to respond. As such, core information can be constricted by both structure and context (Comfort et al. 2004).

Hierarchy: Communication can fail in a disaster climate due to rigid hierarchical systems. "In rigidly hierarchical systems, there are overt barriers to the free flow of information, even when that information is of a kind that is crucial for effective managerial decision-making" (Weir 1996: 119). These overt barriers can manifest due to a system's reliance on traditional chain of command (C) structures where decision making is centralised and the leader in charge is effectively responsible for:

> using available resources, planning the employment of, organizing, directing, coordinating, and controlling [military forces] for the accomplishment of assigned missions. It also includes the responsibility for health, welfare, morale, and discipline of assigned personnel (Joint Chiefs of Staff Publication 2003; quoted in Alberts and Hayes 2003: 14).

Such a system creates a 'mind-set' which disables imagination, dynamism and foresight (Masys et al. 2014; Weick 1990). Rigid hierarchical systems also enable the possibility of core information either evaporating or getting delayed in the structures of human built organisations. For example, in the case of Fukushima nuclear accident, The National DIET of Japan reported (2012: 38):

> The accident was the result of Tokyo Electric Power Company's (TEPCO) failure in preparing against earthquakes and tsunamis, despite repeated warnings about the potential for such catastrophes. Although TEPCO had reviewed possible countermeasures for the kind of events that subsequently transpired, it postponed putting any measures into place for the other events, using the scientific improbability of such events as an excuse.

Inaccuracy of information: Inaccuracy of the core information can lead to systems failure. This inaccuracy can stem from different frames of reference used by professional groups (Weick 1990; Weir 1996). Professional groups use specialist jargons which are often the outcome of strong vertical divisions between themselves. Some of these divisions between the professional groups are discussed in the next section on world views. Government, non-government and popular media use different forecasting terminology when warning the public and private businesses (GIZ 2012; 2015; Herrmann 2009). These warnings are often confusing. They also differ distinctly from the indigenous early warning systems and practices (GIZ 2012; Herrmann 2009; Romo-Murphy et al. 2011). The early warning systems can also lack in providing "a reliable warning of a potential threat with enough lead-time for recipients to take appropriate evasive action" (Herrmann 2009: 14). In such a context, the "problem with early warnings boils down to the common difficulty of perception versus reality" (Herrmann 2009: 14).

Of equal importance is the matter of 'trust' (Morgner 2013) between the secondary organisations who are generating and cascading the core information and the end users. If the core information generated is often inaccurate, this will lead to mistrust in the system and potentially fatal consequences. Also, communication that does not specify the actions required to save lives will be futile. Likewise, a response without core information will be blind in response. Both of these problems can cause deaths and systems failure in a disaster climate.

2.6.3 World Views

Systems can also fail due to conflicting world views. Under the aegis of UN's disaster risk reduction (DRR) framework, disaster management is a combination of early warning practitioners, Category 1 and 2 responders, gender and disaster risk reduction specialists at international, national and local levels—to mention a few. These actors have their own world views or *Weltanschauung* as how best to avert disaster risks. Some of the world views of the disaster risk reduction, vulnerability and gender studies were discussed earlier in this chapter in order to explain why deaths occur. Strategies and targets emanate from the world views that these actors adopt. As such, examining the world views of actors is extremely important in this research.

A world view is a "complex set of perceptions, attitudes, values and motivations that characterise an individual or group" (Waring 1989: 12). It is also a kind of perceptual 'window' or 'tinted spectacles' through which each of us interprets the world (Checkland 1981; Waring 1989). It encapsulates the "notion that our experiences of the world are mediated or interpreted in terms of our purposes, knowledge, values, and expectations etc., which have developed in particular ways through our previous experiences" (Mingers 1980: 6). Past experiences and world views shape mental models (Mingers 1980) which could either be detrimental (as we saw in the case of Fukushima) or an enabler for an organisation to learn (Senge 1990).

It is argued here that the problems of coordination and communication are tightly coupled with the conflicting world views of the different actors and organisations of the disaster management system. One of the consequences of these subjective world views is a lack of an 'overall objective' or a 'goal' (Jenkins 1969) to avoid deaths. In the context of this research, lack of an objective is rather a lack of input into the system to reduce 'deaths' and this is demonstrated below.

Category 1, 2 and Early Warning Practitioners: Currently the generation and dissemination of core information by meteorologists is done in a gender neutral way. They do not target 'at-risk' community as their primary end user. Instead their target group includes decision makers (Category 1 and 2), media, businesses such as insurance companies, the aviation sector and grain producers—to mention a few (Glantz 2009). The purpose is to communicate effective warnings for current and future threats. This research in Odisha had similar findings. However, some discretion is left with the Category 1 and 2 responders to decide who they want to warn and how—as observed in the case of Odisha which will be discussed in the subsequent chapters.

Disaster Risk Reduction Advocates: Much like its predecessor the Hyogo Framework for Action, the Sendai Framework emphasises the importance of early warnings 'that are people centred' (UN 2005: 4–5, iid.9; UN 2015; UNISDR 2006). People centred approaches are highly useful but at the same time the concept of 'people' raised by the UN requires further examination. Emphasis on people or humans has the potential to assume gender-neutrality, which can "often be an expression of the masculine in which the gender dimension can be overlooked, hence providing only a partial understanding of [human security] issues", according to Hudson (2005: 157).

People centred approaches may also be dubbed as the 'whole community approach' or 'risk reduction approach' and they can often overlook the needs of vulnerable groups, including women and men during disasters (Ikeda 2009; Ray-Bennett 2016a). In this light, this can be considered as an *omission* in the disaster risk reduction framework. Omissions are the by-product of latent errors. They are "the failure to carry out some of the actions necessary to achieve a desired goal" (Reason 1990: 184).

Gender and Disaster Advocates: In January 2005, the Hyogo Framework mainstreamed gender into all "disaster risk management policies, plans and decision making processes, including those related to risk assessment, early warning, information management, education and training". More concretely it mainstreamed gender into two Priorities for Action including Early Warning (Priority 2) and Knowledge Management and Education (Priority 3) (UN 2005). This event is considered as a hallmark for the gender and disaster advocates. It also marked a process of converging two world views—one of the disaster risk reduction community and the other of the gender and disaster community. Currently, there is a paucity of research in assessing whether gender mainstreaming has actually succeeded in reducing men and women's disaster risks and vulnerability.

Gender mainstreaming studies, in general, pursue two agendas. The first agenda is integrating a gender approach into existing policies/programmes through gender equality and equity. This is known as integrationist approach. The second agenda is to assess women's empowerment inside and outside an organisation through structural change. This is known as transformative or agenda setting approach (Jahan 1995; Moser and Moser 2005; Porter and Sweetman 2005; Riley 2004; Tiessen 2004). Nevertheless, this approach has limitation in the context of this research. Although the current agendas (equality/equity and empowerment) are highly important in everyday life of the poor and vulnerable, they lack in prioritising the agenda of avoiding deaths of the vulnerable groups in disasters. As such, they too commit a latent error of not prioritising the agenda of directly avoiding women, men and children's deaths in disasters.

Having presented the framework for systems failure, the next two chapters present the case of the Super-Cyclone of 1999 and Cyclone Phailin of 2013. In doing so, the analytical advantages of systems failure are discussed.

References

Aalst MV, Kellett J, Pichon F, Mitchell T (2013) Incentives in disaster risk management and humanitarian response: background note for world development report. Overseas Development Institute, London

Acton JM, Hibbs M (2012) Why Fukushima was preventable. The Carnegie Papers. http://carnegieendowment.org/files/fukushima.pdf. Accessed 6 Jan 2014

Alberts DS, Hayes RE (2003) Power to the edge: Command, control in the information age. CCRP Publication Series. www.dodccrp.org. Accessed 6 Feb 2014

Alexander D (1993) Natural disasters. UCL Press, London

Anderson H, Goolishian HA (1988) Human systems are linguistic systems: preliminary and evolving ideas about the implications for clinical theory. Fam Process 27:371–393. doi:10.1111/j.1545-5300.1988.00371.x

Ariyabandu MM (2009) Sex, gender and gender relations in disasters. In: Enarson E, Chakrabarti PGD (eds) Women, gender and disaster: global issues and initiatives. Sage Publications, New Delhi

Ariyabandu MM, Wicramasinghe M (2003) Gender dimensions in disaster management: a guide for South Asia. ITDG South Asia Publication, Colombo

Arnoldi J (2001) Niklas Luhmann: an introduction. Theory Cult Soc 18(1):1–3. doi:10.1177/02632760122051607

Aus Aid (2014) Women's and children's health in emergencies. Aus Aid. www.ausaid.gov.au. Accessed 17 Feb 2014

Bacon CM, Sundstrom W, Stewart IT, Beezer D (2017) Vulnerability to cumulative hazards: Coping with the coffee leaf rust outbreak, drought, and food insecurity in Nicaragua. World Dev. 93(May):136–152. http://dx.doi.org/10.1016/j.worlddev.2016.12.025

Bankoff G (2001) Rendering the world unsafe: "Vulnerability" as western discourse. Disasters 25 (1):19–35. doi:10.1111/1467-7717.00159

Barnshaw J, Trainor J (2007) Race, class and capital. In: Brunsma DL, Overfelt D, Picou JS (eds) The sociology of Katrina: perspectives on a modern catastrophe. Rowman and Littlefield Publishers Inc, New York, pp 91–106

BBC (2014) Zeebrugge disaster: Seaford's Jane Hind tells of family loss, 18 Aug. http://www.bbc.co.uk/news/uk-england-28785434. Accessed 4 Jan 2016

Blockley DI (1996) Hazard engineering. In: Hood C, Jones C (eds) Accident and design: contemporary debates in risk management. UCL Press, Oxon, pp 31–39

Boehnert H (2009) Geographic information systems. In: Glantz MH (ed) Heads up!: early warning systems for climate-water-and weather-related hazards. United Nations University Press, Paris, pp 7–10

Bradshaw S, Fordham M (2013) Women, girls and disasters: a review for DFID. Department for International Development, London

Broughton E (2005) The Bhopal disasters and its aftermath: a review. Environ Health 4(2005):4–6. doi:10.1186/1476-069X-4-6

Bryant E (2005) Natural hazards, 2nd edn. Cambridge University Press, Cambridge

Bunge M (1977) General systems and holism. Gen Syst 22:87–90

Burton B, Kates RW, White GF (1993) The environment as hazard, 2nd edn. The Guildford Press, New York

Cannon T (1994) Vulnerability analysis and the explanation of "natural disasters". In: Varley A (ed) Disasters, development and environment. Wiley, New York, pp 14–30

Carter WN (1999) Disaster management: a disaster manager's handbook. Asian Development Bank, Manila

Chakravarti U (2003) Gendering caste: through a feminist lens. Stree, Calcutta

Chambers R (1989) Editorial introduction: vulnerability, coping and policy. IDS Bull 20(2):1–7. doi:10.1111/j.1759-5436.2006.tb00284.x

Channel 4 (2015) Tsunami: ten years. ITV broadcast on 10 January 2015

Chatterjee A, Gupta D, Jain N (2010) Coordination of disaster response: potential and challenges from Indian experiences. Knowledge Communication on Children in India. http://www.redr.org.in/uploads/Coordination_of_Disaster_Response.pdf. Accessed 10 Feb 2015

Checkland P (1981) Systems thinking: systems practice. Wiley, Chichester

Checkland P (1985) From optimizing to learning: a development of systems thinking in the 1990s. J Oper Res Soc 36(9):757–767. doi:10.1057/jors.1985.141

Comfort LK, Ko K, Zagoreck A (2004) Coordination in rapidly evolving disaster response systems: the role of information. Am Behav Sci 48(3):295–313. doi:10.1177/000276420 4268987

Davis I, Gupta SP (1991) Technical background paper. In: Bank Asian Development (ed) Disaster mitigation in Asia and the Pacific. Asian Development Bank, Manila

Dekker S (2003) Accidents are normal and human error does not exist: a new look at the creation of occupational safety. Int J Occup Saf 9(2):211–218. doi:10.1080/10803548.2003.11076564

Dekker S (2006) The field guide to understanding human error. Ashgate, Surrey

de Senarclens P (1997) How the UN promotes development through technical assistance. In: Rahnema M, Bawtree V (eds) The post-development reader. Zed Books, London, pp p190–p201

DFID (2013) Helping developing countries deal with humanitarian emergencies. DFI, London

Dilley M, Chen RC, Deichmann U (2005) Disaster risk management series. No. 5. Natural disaster hotspots: a global risk analysis. The World Bank Publication, Washington DC

Douglas M (1966) Purity and danger: an analysis of concept of pollution and taboo. Routledge, London

Eklund L, Tellier S (2012) Gender and international crises response: do we have the data, and does it matter? Disasters 36(4):589–608. doi:10.1111/j.1467-7717.2012.01276.x

Emery F (1993) Characteristics of socio-technical systems. In: Trist E, Murray H (eds) The social engagement of social science, vol II. The socio-technical erspective. University of Pennsylvania Press, Philadelphia, pp 157–186

Fennell D (1988) Investigation into the Kings Cross Underground Fire. Department of Transport, Her Majesty's Stationary, London. Available online at: http://www.railwaysarchive.co.uk/documents/DoT_KX1987.pdf. Accessed 4 January 2016
Garrison RW, Kirzner I (1987) Friedrich August Von Hayek. In: Eatwell J (ed) The new Palgrave: a dictionary of economics. Macmillan Ltd, London, pp 609–614
GENI (2014) Multilateral organizations. Global Energy Network Institute. http://www.geni.org/globalenergy/library/organizations/index.shtml#global. Accessed 29 Jan 2015
Gilbert C (1998) Studying disaster: changes in the main conceptual tools. In: Quarantelli EL (ed) What is a disaster? Perspectives on the questions. Routledge, London, pp 11–18
Gillies J (2014) Scientists sound alarm on climate change. New York Times, March 18. http://www.nytimes.com/2014/03/18/science/scientists-sound-alarm-on-climate.html?_r=0. Accessed 18 Mar 2014
GIZ (2012) LFEWS: local flood and early warning system. GIZ Publication, Manila
GIZ (2015) Harmonisation of early warning alert levels. GIZ Publication, Manila
Glantz M (2009) Introduction. In: Glantz M (ed) Heads up!: early warning systems for climate, water and weather-related hazards. United Nations University Press, New York, pp 1–28
Global Emergency Group (undated) IASC real-time evaluation of the humanitarian response to the Horn of Africa drought crisis, Kenya. Interagency Standing Committee, Virginia
GoO (2014a) District disaster management plan 2014–15: Puri District. Emergency Operation Centre, Puri
GoO (2014b) Ganjam disaster district management plan 2014–15. Chatrapur District Emergency Operation Center, Chatrapur
Grey C (2009) A very short fairly interesting and reasonably cheap book about studying organizations, 2nd edn. Sage Publication Ltd, London
Grint K (2008) Wicked problems and clumsy solutions: the role of leadership. Clin Leader I (II):11–15
Hanai T (2014) Landslides hit Japan's Hiroshima, killing at least 36. Reuters, Aug 21. http://www.reuters.com/article/us-japan-landslides-idUSKBN0GK06720140822. Accessed 12 July 2015
Haque E, Zaman M (1994) Vulnerability and responses to riverine hazards in Bangladesh: a critique of flood control and mitigation approaches. In: Varley A (ed) Disasters, development and environment. Wiley, New York, pp 65–79
Haque U, Hashizume M, Kolivras KN, Overgaard HJ, Das B, Yamamoto T (2011) Reduced death rates from cyclones in Bangladesh: what more needs to be done?. Bull World Health Organ 2012(9):150–156. doi:10.2471/BLT.11.088302
Harun-Al-Rashid AKM (1997) The cyclone preparedness programme in Bangladesh. Community based approach in disaster management-I. ADPC, Bangkok
Herrmann S (2009) Remote sensing. In: Glantz M (ed) Heads up!: early warning systems for climate, water and weather-related hazards. United Nations University Press, New York, pp 10–11
Hewitt K (2013) Environmental disasters in social context: toward a preventive and precautionary approach. Nat Hazards 66(1):3–14. doi:10.1007/s11069-012-0205-6
Horlick-Jones T (1996) The problem of blame. In: Hood C, Jones C (eds) Accident and design: contemporary debates in risk management. UCL Press, Oxon, pp 61–71
Hudson H (2005) 'Doing' security as though humans matter: a feminist perspective on gender and the policies of human security. Secur Dialogue 36(2):153–174. doi:10.1177/0967010605054642
Ikeda K (2009) How women's concerns are shaped in community based disaster risk management in Bangladesh. Contemp South Asia 17(1):65–78. doi:10.1080/09584930802624679
Integrated Research on Disaster Risk (2015) Guidelines on measuring losses from disasters: Human and economic impact indicators. Integrated Research on Disaster Risk, Beijing
IPCC (2007) Climate change 2007: synthesis report. IPCC, Switzerland
IPCC (2012) Managing the risks of extreme events and disasters to advance climate change adaptation. In: Field CB, Barros V, Stocker TF, Qin D, Dokken DJ, Ebi KL, Mastrandrea MD, Mach KJ, Plattner JK, Allen SK, Tignor M, Midgley PM (eds) A special report of working

Groups I and II of the intergovernmental panel on climate change. Cambridge University Press, Cambridge
Jahan R (1995) The elusive agenda: mainstreaming women in development. Zed Books, London
Jaspars S, Shoham J (1999) Targeting the vulnerable: a review of the necessity and feasibility of targeting vulnerable households. Disasters 23(4):359–372. doi:10.1111/1467-7717.00125
Jenkins GM (1969) The systems approach. J Syst Eng 1:3–49
Kaiser R, Spiegel PB, Henderson AK, Gerber ML (2003) The application of geographic information systems and global positioning systems in humanitarian emergencies: lessons learned, programme implications and future research. Disasters 27(2):127–140. doi:10.1111/1467-7717.00224
Kalsi SR (2003) Monitoring and forecasting of tropical cyclones and their associated effects. In: Gupta HK (ed) Disaster management. Universities Press, Mumbai, pp 48–84
Kapur A (2010) Vulnerable India: a geographical study of disasters. Sage Publications, New Delhi
Kast FE, Rosenzweig JE (1972) General systems theory: applications for organization and management. Acad Manag J 15(4):447–465. doi:10.2307/255141
Kolen B, Helsloot I (2012) Time needed to evacuate the Netherlands in the event of large scale flooding: strategies and consequences. Disasters 36(4):700–722. doi:10.1111/j.1467-7717.2012.01278.x
Lalonde C (2011) Managing crisis through organisational development: a conceptual framework. Disasters 35(2):443–464. doi:10.1111/j.1467-7717.2010.01223.x
Lass W, Haas A, Hinkel J, Jaeger C (2011) Avoiding the avoidable: towards a European heat waves risk governance. Int J Disaster Risk Sci 2(1):1–14. doi:10.1007/s13753-011-0001-z
Latour B (2005) Reassembling the social: an introduction to actor-network theory. Oxford University Press, Oxford
Luhmann N (1993) Risk: a sociological theory (trans. Barrett R). Walter de Gruyter & Co, Berlin
Luhmann N (1999) Introduction to systems theory (trans. Giligen P). Polity, Cambridge
Masys AJ, Ray-Bennett NS, Shiroshita H, Jackson P (2014) High impact/low frequency extreme events: enabling reflection and resilience in a hyper-connected world. Proc Econ Finan 18 (2014):772–779. doi:10.1016/S2212-5671(14)01001-6
McNeill P, Chapman S (1985) Research methods, 3rd edn. Routledge, Oxon
Midgley G (ed) (2003) Systems thinking: Volume 1: General Systems Theory, Cybernetics and Complexity. London: Sage
Midgley G (2014) An introduction to systems thinking for tackling wicked problems. M.Sc. in Risk, Crisis and Disaster Management Study School, University of Leicester, 29 Sept
Miles R (2012) Empowering emergency responders. In: Bennett S (ed) Innovative thinking in risk, crisis, and disaster management. Gower, Surrey, pp 25–40
Mingers JC (1980) Towards an appropriate social theory for applied systems thinking: critical theory and soft systems methodology. J Appl Syst Anal 7:41–49
Modh RS (2010) Managing natural disasters: hydrological, marine and geological disasters. Macmillan, Delhi
Moore R, Verity A (2014) Hashtag standards for emergencies: OCHA policy and studies brief. United Nations Office for the Coordination of Humanitarian Affairs (OCHA), Geneva
Morgner C (2013) Trust and confidence: History, theory and socio-political implications. Hum Stud 36(4):509–532. doi:10.1007/s10746-013-9281-1
Moser CON, Moser A (2005) Gender mainstreaming since Beijing: a review of success and limitation in international institutions. Gender Dev 13(2):11–22. doi:10.1080/135520705123 31332283
Nathan F (2009) Natural disasters, vulnerability and human security: Facing global environmental change. Hexagon Ser Human Environ Secur 4(Part II):103–118. doi:10.1007/978-3-540-68488-6_87
Neumayer E, Plümper T (2007) The gendered nature of natural disasters: the impact of catastrophic events on the gender gap in life expectancy, 1981–2002. Ann Assoc Am Geogr 97 (3):551–566. doi:10.1111/j.1467-8306.2007.00563.x

Perrow C (1999) Normal accidents: living with high-risk technologies, 2nd edn. Princeton University Press, Princeton
Porter F, Sweetman C (2005) Editorial. Gender Dev 13(2):2–10. doi:10.1080/13552070512331332282
Prizza R (2007) The role of coordination in disaster management. In: Pinkowski J (ed) Disaster management handbook. CRC Press, New York, pp 75–96
Quarantelli EL (1977) Disasters: theory and research. Sage Publications, London
Rashid FS, Michand S (2000) Female adolescents and their sexuality: notions of honour, shame, purity and pollution during the floods. Disasters 24(1):54–70. doi:10.1111/1467-7717.00131
Ray-Bennett NS (2009a) Caste, class and gender in multiple disasters: the experiences of women-headed households in an Oriya village. VDM Verlag, Saarbrucken
Ray-Bennett NS (2009b) Multiple disasters and policy responses in pre and post-independence Orissa, India. Disasters 33(2):275–290. doi:10.1111/j.1467-7717.2008.01074.x
Ray-Bennett NS (2009c) The influence of caste, class and gender in surviving multiple disasters in Orissa, India. Environ Hazards 8(1):5–22
Ray-Bennett NS (2010) Multiple disasters and micro-credit: can it reduce women's vulnerability?: the case of Tarasahi in Orissa, India. Disasters 34(1):240–260. doi:10.1111/j.1467-7717.2009.01127.x
Ray-Bennett NS (2016) Exploring the meaning of securitisation for "gender and disaster". In: Racioppi L, Rajagopalan S (eds) Women and disasters in South Asia: survival, security and development. Routledge, New York, pp 277–301
Reason J (1990) Human error. Cambridge University Press, Cambridge
Reason J (1997) Managing the risks of organisational accidents. Ashgate, Aldershot
Riley J (2004) Some reflections on gender mainstreaming and intersectionality. Dev Bull 66:82–86
Rittel HWJ, Webber MM (1973) Planning problems are wicked problems. Policy Sci 4(1973): 155–169. In: Cross N (ed) Developments in design methodologies. Wiley, Chichester
Romo-Murphy E, James R, Adams M (2011) Facilitating disaster preparedness through local radio broadcasting. Disasters 35(4):801–815. doi:10.1111/j.1467-7717.2011.01234.x
Samal KC, Meher S, Panigrahi N (2003) Beyond relief disaster mitigation, livelihood rehabilitation and the post-cyclone recovery in Orissa: village level studies in three most cyclone affected districts in Orissa. Nabo Krishna Centre for Development Studies Publication, Bhubaneswar
Sen AK (2003) Forward. In: Farmer P (ed) Pathologies of power: health, human rights, and the new war on the poor. University of California Press, California
Senge P (1990) The leader's new work: building learning organizations. Sloan Manag Rev 32 (1990):7–23
Sivakumar MVK (2005) Impacts of natural disasters in agriculture, rangeland and forestry: an overview. In: Sivakumar MVK, Motha RP, Das HP (eds) Natural disasters and extreme events in agriculture: impacts and mitigation. Springer, New York, pp 1–22
Smith K, Petley DN (2009) Environmental hazards: assessing risk and reducing disaster, 5th edn. Routledge, New York
Srivastava SK (2009) Making a technological choice for disaster management and poverty alleviation. Disasters 33(1):58–81. doi:10.1111/j.1467-7717.2008.01062.x
Tanaka A (2015) Toward a theory of human security. JICA-R1 Working Paper, No 91. JICA Research Institute Publication, Tokyo
The National Diet of Japan (2012) The official report of the Fukushima nuclear accident independent investigation commission. Fukushima Nuclear Accident Independent Investigation Commission (NAIIC), Japan, Tokyo
Tiessen R (2004) Re-inventing the gendered organization: staff attitudes towards women and gender mainstreaming in NGOs in Malawi. Gender Work Organ 11(6):689–708. doi:10.1111/j.1468-0432.2004.00255.x
Toft B, Reynolds S (2005) Learning from disasters: a management approach, 3rd edn. Palgrave Macmillan, Hampshire

TrustOrg (2013) Women in disasters. Thomson Reuters Foundation. http://www.trust.org/spotlight/Women-the-poorer-half-of-the-world?tab=briefing. Accessed 17 Feb 2014
UN (2005) Hyogo framework for action 2005–2015: building the resilience of nations and communities to disasters. In: World conference on disaster reduction, Kobe, Hyogo, Japan, 18–22 Jan 2005
UN (2015) Sendai framework for disaster risk reduction 2015–2030. UNISDR, Geneva. http://www.unisdr.org/we/coordinate/sendai-framework. Accessed 24 June 2016
UNDP (2007) Human development report 2007/08. Fighting climate change: human solidarity in a divided world. Palgrave Macmillan, New York
UNISDR (2006) Developing early warning systems, a checklist. In: Third international conference on early warning. http://www.unisdr.org/we/inform/publications/608. Accessed 14 Jan 2014
UNISDR (2009a) 2009 UNISDR terminology on disaster risk reduction. UNISDR, Geneva. http://www.preventionweb.net/files/7817_UNISDRTerminologyEnglish.pdf. Accessed 2 Aug 2016
UNISDR (2009b) Making disaster risk reduction gender-sensitive: policy and practical guidelines. UNISDR, Geneva. http://www.unisdr.org/we/inform/publications/9922. Accessed 4 May 2012
UNISDR (2015a) Making development sustainable: the future of disaster risk management: Global assessment report on disaster risk reduction 2015. UNISDR, Geneva
UNISDR (2015b) Japan provides early warning example: UN world conference in disaster risk reduction 2015 Sendai Japan. In: Third UN world conference on disaster risk reduction. www.wcdrr.org. Accessed 16 Mar 2015
Varley A (1994) The exceptional and the everyday: vulnerability analysis in the international decade for natural disasters reduction. In: Varley A (ed) Disasters, development and environment. Wiley, New York, pp 1–11
Vickers G (1983) Human systems are different. J Appl Syst Anal 10(1983):3–13
Walker C, Broderick J (2006) The civil contingencies act 2004: risk, resilience and the law in the United Kingdom. Oxford University Press, Oxford
Waring A (1989) Systems methods for managers. Blackwell Scientific Publications, Oxford
Weick KE (1990) The vulnerable systems: an analysis of the Tenerife air disaster. J Manag 16 (3):571–593. doi:10.1177/014920639001600304
Weir DTH (1996) Risk and disasters: The role of communications breakdown in plane crashes and business failure. In: Hood C, Jones C (eds) Accident and design: contemporary debates in risk management. UCL Press, Oxon, pp 114–126
Wenger E (1998) Communities of practices: learning, meaning, and identity. Cambridge University Press, New York
WHO (2002) Gender and health in natural disasters. WHO: gender, equity, human rights. http://www.who.int/gender/other_health/en/genderdisasters.pdf. Accessed 9 December 2014
Winchester P (2000) Cyclone mitigation, resource allocation and post-disaster reconstruction in south India: lessons learnt from two decades of research. Disasters 24(2):18–37
Wisner B, Blaikie P, Cannon T, Davis I (2004) At risk: natural hazards, people's vulnerability and disasters, 2nd edn. Routledge, London
Zaman MQ (1999) Vulnerability, disaster, and survival in Bangladesh: Three case studies. In: Smith O, Hoffman S (eds) The angry earth: disaster in anthropological perspective. Routledge, New York, pp 192–212

Chapter 3
Super-Cyclone in 1999

Abstract This chapter introduces the case studies. On 17–18 October 1999, Odisha, an Indian state was affected by a super-cyclone (wind velocity of 270–300 km/h) which killed more than 10,000 people. In 2013, Odisha was hit by another cyclone called Phailin on 12 October 2013 (wind velocity 260 km/h). Only 86 people died. This chapter presents the gender disaggregated death data from the four districts which were severely hit in 1999. It also provides the narratives of seven elite respondents who included the Director of the Indian Meteorology Department, Regional Director of UNDP, Manager of Odisha State Disaster Management Authority, Deputy Relief Commissioner of the Special Relief Organisation, as well as three Emergency Officers from Ganjam, Jagatsinghpur and Puri districts. Systems failure provides a novel perspective to analyse the reasons why there were so many deaths in the Super-Cyclone.

Keywords Case studies · Super-Cyclone · Cyclone Phailin · Deaths · Ganjam · Jagatsinghpur · Puri · Cuttack · Coordination failure · Systems failure · World views

Jagatsinghpur, Cuttack, Puri and Ganjam were the sampled districts. These districts were selected because they experienced the Super-Cyclone and Cyclone Phailin. *The district of Jagatsinghpur* is surrounded by Cuttack district in the west, Kendrapara in the east and Puri in the south (see the political map of Odisha in Chap. 1). According to the 2011 census the total population of Jagatsinghpur was 1,136,971. Of this 577,865 were male and 559,106 were female (Census of India 2011). Jagatsinghpur has eight blocks, 194 gram panchayats, 1320 revenue villages and two municipalities (GoO 2013). The total geographic location of Jagatsinghpur is 1759 km^2 and it is 54 km from Bhubaneswar (the capital of Odisha). The economy of Jagatsinghpur is based on agriculture, small businesses, cattle rearing, betel vines, prawn cultivation and fishing (GoO 2013; Ray-Bennett 2009).

The district of Ganjam is surrounded by Puri, Khurda and Nayagarh districts in the east and Kandhamal and Gajapati districts in the west. The Bay of Bengal touches the eastern frontier of this district extending a coastline of 60 km. Rushikulya, Dhanei,

N.S. Ray-Bennett *Avoidable Deaths*, SpringerBriefs in Environmental Science,
https://doi.org/10.1007/978-3-319-66951-9_3

Bahuda, and Ghoda Hada are the major rivers in this region. Together they support the fishing, agriculture and the generation of electricity for this district. The area of this district is 9499.7 km^2. According to the census of 2011, the total population of Ganjam was 3,529,031, of which 1,779,218 were male and 1,749,813 were female. It is the most populous district of Odisha (Census of India 2011). Ganjam has 22 blocks, 475 gram panchayats and 3229 revenue villages (GoO 2014b).

The district of Cuttack is surrounded by Kendrapara and Jajpur in the east, Dhenkanal and Angul in the north and north-west and Nayagarh and Khurda in the west. Cuttack "is a narrow strip of land spreading from east to west" (GoO 2008: 6). Topographically, Cuttack is divided into the hilly terrain on the west and Mahanadi delta plain on the east. The delta plains of Mahanadi and Kathjodi are the most fertile lands but are also very low lying and get submerged regularly during monsoon season causing severe floods. The economy of the district is largely based on agriculture. According to the census of 2011, the total population of Cuttack was 2,624,470—of which 1,352,760 were male and 1,271,710 were female. Cuttack is the second largest populated district of Odisha after Ganjam. Cuttack has 14 blocks, 342 gram panchayats and 1950 villages spread over 3932 km^2. Cuttack was separated from Jagatsinghpur in 1993. It is about 30 km from Bhubaneswar. The vast network of irrigation canals established by the British government after the Great Famine of 1866 is still the backbone of agriculture. Cuttack city is the commercial capital of Odisha because it controls *Malgodowns* of the state (GoO 2008). Due to their geographic location, these districts experience recurrent floods, cyclones, heat waves, droughts, hail storms and lightning.

Due to their geographic location, the sampled districts experience recurrent floods, cyclones, heat waves, droughts, hail storms and lightning. Table 3.1 provides the detail of the deaths caused in these disasters from 1999 to 2013.

Since 2003, the data suggests that there has been an increase in deaths due to lightning. This was an interesting finding. Lightning deaths attract less attention from the media and government, due to their isolated nature. Although the focus of this research was on the Super-Cyclone and Cyclone Phailin, the author explored if there was any comparable data on lightning prior to 1999. According to the Deputy Relief Commissioner, data prior to 1999 was either not maintained or has been lost due to the Super-Cyclone. Hence, it was difficult to establish whether deaths due to lightning after 1999 are a new phenomenon or something that had always been there but remained undetected. All the seven respondents acknowledged the increasing nature of this risk on humans, especially in rural areas, but as a hazard it is still not considered of sufficient importance to implement mitigation measures. This 'new risk' evades public and media attention because they cause less physical devastation compared to flooding and cyclones and the impact caused is geographically scattered, sometimes only causing the death of a single person at a time. Nevertheless, the Deputy Relief Commissioner of the Special Relief Organisation agreed to look into the matter of lightening (interview with Rajiv Bhawan held on 22 July 2014).

Table 3.1 Disasters and deaths in the sampled districts from 1999 to 2013 (Compiled from the White Papers and Reports produced by the District Emergency Cell, Ganjam and State Relief Organisation, Bhubaneswar)

Year	Number of deaths by hazard					
	Cyclone	Flood	Drought	Hail storm	Lightning	Heat wave
1999	10,086 (Super Cyclone)	–	–	–	–	91 (DWDNA)
2000	–	–	–	–	–	29
2001	–	102 (DWDNA)	–	–	–	26
2002	–	–	–	3	–	41 (DWDNA)
2003	–	32 (DWDNA)	–	–	222	67
2004	–	10	–	–	274	43
2005	–	13	–	1	245	199
2006	–	105	–	2	204	21
2007	–	59	–	6	331	44 (DWDNA)
2008	–	110	–	20	362	68
2009	–	59	–	5	214	89 (DWDNA)
2010	–	14 (DWDNA)	–	14	227	100
2011	–	7	–	5	360	22
2012	–	–	–	–	–	–
2013	86 (Phailin)	37	–	–	–	–

DWDNA District wise data not available

3.1 Super-Cyclone and Cyclone Phailin

In October 1999, Odisha was hit by two cyclones within a period of two weeks. The first cyclone occurred on 17–18 October affecting two coastal districts: Ganjam and Gajapati, and the later cyclone on 29–30 October affecting 13 coastal districts. Whilst the district authorities and affected communities were recovering from the extensive damages caused by the first cyclone, the meteorologists forecasted another 'very severe cyclone' for the coastal districts. The Indian Meteorological Department, later, termed the second cyclone as a 'Super-Cyclone', due to its high wind velocity of 270–300 km/h, its unprecedented storm surge which was 5–7 m high and the torrential rainfall over 48 h that caused devastating floods in the major river basins. The intensity of the cyclone was such that it killed 10,086 people, caused severe socio-economic devastation and activated the Orissa Relief Code (the then sole disaster policy document for the state). It also put Odisha in the spotlight internationally because the super-cyclone coincided with the tail-end of the UN's International Decade for Disaster Reduction (IDNDR) (1990–2000). Consequently, there was an overwhelming post-disaster response by the state, neighbouring states,

NGOs, INGOs and multinational organisations (Behera and Sarkar 2003; GoO 2002a, b, 2005; Ray-Bennett 2009; Samal 2003).

According to Swain et al. (2006: 4), nearly "18.38 lakh [1,838,000] houses were damaged, of which 6.84 lakh [684,000] houses were fully collapsed and 11.27 lakh [1,127,000] houses were partially damaged. About 27,479 houses were washed away by the tidal waves during the super cyclone, of which 12,693 houses were from Jagatsinghpur district and 13,821 houses were from Balasore. The physical infrastructures including school and college buildings, government offices were severely damaged". The Super-Cyclone also affected the crops and inundated 1,733,000 ha of agricultural land by sea water. About 15,681,000 people were affected by the Super-Cyclone, which meant 42.7% of the total population of Odisha at that time (Swain et al. 2006).

Fourteen years after the Super-Cyclone, Odisha was hit by another cyclone called Phailin on 12 October 2013. According to Haeseler (2013: 2–8):

> Cyclone PHAILIN originated from an upper low across the Gulf of Thailand on 5 October 2013. On the 8th it became apparent as a depression in the surface pressure field across the Bay of Bengal, and overnight 8th/9th [...] wind speeds [reached] more than 34 kn (63 km/h) it was designated as "tropical storm". [...] Apart from a temporary weakening during the crossing of the Andaman Islands, PHAILIN strengthened steadily and was declared as "cyclone" on the 10th. PHAILIN reached its maximum 1-minute average wind speeds on 11th/12th with 140 kn (260 km/h). This is comparable to a category 5 hurricane. [...].
>
> The cyclone slightly weakened as it approached the Indian east coast. On 12 October 2013 at 06:00 UTC, PHAILIN had a diameter of approximately 500 km, being within the average size of tropical cyclones over the north Indian Ocean. At 12 UTC, strong winds with wind speeds of more than 35 kn (65 km/h) already affected large coastal areas though the centre of the storm had not reached the Indian coast. [...]
>
> PHAILIN made landfall [...] near the city of Gopalpur on 12 October at about 20:30 to 21:30 local time [...]. Already hours before the landfall, the coastal regions were affected by storm and heavy rains. [...] The wind speed reached up to 25 m/s (90 km/h), and gusts of about 36 m/s (130 km/h). [...]. From 11 to 13 October, Paradeep recorded a noticeable storm surge[1] [...] of 1.5 m. [...] the storm surge, heavy rains caused extensive floods."

Phailin was dubbed as a 'very severe cyclonic storm' (GoO 2014b) which affected about 13,200,000 people (Mishra 2013), of which 983,642 were evacuated by the district and local authorities (Bisoi 2013). 18 districts, 171 blocks, 2,164 gram panchayats, and 18,374 villages were affected by Phailin (Mishra 2013). More than 201,652 mud houses and 7,604 concrete houses were damaged.

However, compared to the Super-Cyclone of 1999, Phailin was less intense in three aspects. The wind velocity of the Super-Cyclone reached up to 300 km/h compared to 260 km/h in Phailin (Haeseler 2013). Second, according to the United Nations Environment Programme, the storm surge reached 3.5 m in the coastal regions compared to 5 m to 6 m during the Super-Cyclone (Harriman 2013). Third,

---

[1] "The difference between observed water level and predicted tidal water level is called the height of the storm surge" (Haeseler 2013: 7).

a 24-h precipitation total of 169 mm was recorded on 13 October 2013, whereas a 24-h precipitation total of about 520 mm was recorded at the weather station of Paradeep on 30 October 1999 (Haeseler 2013). Although the anatomy of these two tropical cyclones was different, they are comparable on two grounds: they tested the disaster management systems of Odisha to their limits; and they presented a 'window of opportunity' to assess the strengths and limitations of the systems built by the government and non-government organisations between 1999 and 2013 to save lives.

3.2 Systems Failure in 1999

The Super-Cyclone of 1999 is argued as a case of systems failure. This is because 10,086 people died. In the sampled districts, Cuttack had 456 deaths (256 male and 200 female); Ganjam had 105 deaths (105 male and 78 female); Jagatsinghpur had 8,119 deaths (4,658 male and 3,461 female); and Puri had 301 deaths (170 male and 131 females) (see Table 3.2). Jagatsinghpur had the highest number of deaths. All these deaths were beyond the avoidable death indicators.

Furthermore, Table 3.2 shows that during the Super-Cyclone, the number of male deaths were slightly higher than the female deaths within the sampled districts. There are various reasons why this was the case. For instance, as discussed earlier most men were closer to the coast. To ensure that this data was in proportion to the population size of the district, further analysis was conducted. The figures relating to women's deaths were also analysed. The mortality data was examined to identify any statistically significant variation between the proportion of women who died as a result of disasters and the proportion of women in the general population. Mortality data was collected from the Government of Odisha's district and block offices showing the number of women and men each year who died as a result of cyclones, floods, droughts, hail storms, lightning strikes and heat waves. This data covers the period from 1999 to 2013, the period of time between the Super-Cyclone and Cyclone Phailin. Census data was used to establish the proportion of women in the general population (Table 3.3). As the census data was only collected in 1991,

Table 3.2 Gender disaggregated mortality data in the Super-Cyclone (Compiled from the death records available at the District Emergency Offices in Puri, Jagatsinghpur, Ganjam and Cuttack, 2014)

District	Male deaths	Female deaths	Total
Cuttack	256 (56%)	200 (44%)	456 (100%)
Ganjam	105 (57%)	78 (43%)	183 (100%)
Jagatsinghpur	4658 (57%)	3461 (43%)	8119 (100%)
Puri	170 (56%)	131 (44%)	301 (100%)

Table 3.3 Approximate gender disaggregated population data (rounded to the thousands, 000's) (Compiled from Census of India 1991, 2001, 2011)

District	Gender	1991	2001	2011
Cuttack	Male	1,028 (52%)	1,208 (52%)	1,339 (51%)
	Female	945 (48%)	1,133 (48%)	1,280 (49%)
	Total	1973 (100%)	2341 (100%)	2619 (100%)
Ganjam	Male	1,348 (50%)	1,582 (50%)	1,777 (50%)
	Female	1,356 (50%)	1,579 (50%)	1,743 (50%)
	Total	2704 (100%)	3161 (100%)	3520 (100%)
Jagatsinghpur	Male	513 (51%)	539 (51%)	578 (51%)
	Female	501 (49%)	519 (49%)	559 (49%)
	Total	1014 (100%)	1058 (100%)	1137 (100%)
Puri	Male	663 (51%)	763 (51%)	865 (51%)
	Female	643 (49%)	740 (49%)	833 (49%)
	Total	1306 (100%)	1503 (100%)	1698 (100%)

2001 and 2011, linear interpolation and extrapolation was used to estimate populations for the other years in the period of interest.

The proportion of females in both sets of data was compared using a two-tailed binomial test, the null hypothesis being that there was no difference between the proportions of female deaths as compared to the proportion of females in the general population. Using a level of significance of $\alpha = 0.05$, the null hypothesis was rejected in ten cases (see Table 3.4). In each of these cases, the proportion of fatalities that were female was *less* than the proportion of females in the general population. This suggests that more men died in the sampled locations. Jagatsinghpur had the highest death toll compared to 14 districts owing to its close proximity to Paradeep where the Super-Cyclone made its landfall.

Table 3.4 Statistically Significant Variations Between the Female Proportion of Deaths and the General Population ($p < 0.05$, two-tailed binomial test) (Compiled from Census of India (1991, 2001, 2011); death records compiled from the District Emergency Offices in Puri, Jagatsinghpur, Ganjam and Cuttack, 2014)

District	Year	Number of deaths	Percentage (%) of deaths that were female	Percentage (%) of population that was female	Significance (p)
Cuttack	2003	19	10.5	48.5	0.001
	2005	36	25.0	48.6	0.004
	2006	35	8.6	48.6	<0.001
	2007	37	13.5	48.7	<0.001
Ganjam	2007	22	22.7	48.7	0.018
Jagatsinghpur	1999	8,119	42.6	49.1	<0.001
	2003	13	7.7	49.1	0.003
	2008	28	25.0	49.1	0.013
Puri	1999	301	43.5	49.2	0.049
	2006	9	0.0	49.1	0.004

3.3 Enacting Systems Failure

Interviews with the Director of the Indian Meteorology Department of India, Deputy Relief Commissioner of the Special Relief Organisation, State Programme Officer of United Nations Development Programme, Deputy General Manager of the Odisha State Disaster Management Authority, District Emergency Officers and an Officer at the Revenue and Disaster Department reveal that this high number of human deaths during the Super-Cyclone can be explained due to the failure emanated from very poor coordination, the collapse of communication channels and a world view that lacked a culture of disaster preparedness.

Narratives of the seven participants are produced as they were recorded. However, some minor adjustments relating to English articles, prepositions and tenses were made in order to maintain the flow of the quotes. English was the participants' third language; hence, some adjustments were essential with regard to syntax.

3.3.1 Coordination and Communication Failures

In Chap. 2 it was argued that coordination problems arise when core information is not available for the Category 1 and 2 responders to develop an effective response system. When core information is unavailable, the response system will be blind (Comfort et al. 2004). The State Programme Officer of United Nations Development Programme (interviewed on 23 July 2014, Bhubaneswar) called this blind response a 'charitable response'—one that lies at the mercy and charity of the responders.

According to the Director of the Indian Meteorology Department, coordination of core information failed in Odisha because:

> Prior to 1999 there was no coordination between the government departments. The technology was under-developed. We had to rely on New Delhi and Kolkata for weather forecasts over telephone. There was delay in receiving weather warnings.
>
> (Summarised from author's field diary, meeting held on 21 July 2014, Indian Meteorology Department Office, Bhubaneswar).

According to Harriman (2013), the Indian Meteorology Department was able to generate early warnings for the Super-Cyclone only two days prior compared to four days prior in the case of Phailin. The delay in generating core information affected the decision making processes of local responders in the sampled districts. This is explained in the words of the former District Emergency Officer of Ganjam[2]:

[2]Due to his commendable work in Phailin in the district of Ganjam, he was promoted to the Revenue and Disaster Management Department in Bhubaneswar. Hence, during the author's fieldwork in 2014 he was no longer an Emergency Officer of Ganjam.

> Actually the 1999 Super Cyclone was unexpected for Odisha. It was not expected that a cyclone can hit in such a way that people will die – it was one reason. Second reason, we had less planning that time. We had no plan at that time – just carried out, which are the provisions laid out in the Odisha Relief Code. We didn't have the District Disaster Management Plan, so there was no, you know, authority like OSDMA (Odisha State Disaster Management Authority), no specific person assigned for the duty of monitoring relief and rescue from Bhubaneswar. So, [...] we had less plan and planning, we didn't think that a cyclone can, you know, devastate everything, and people will die like, you know! Yes, this was the major difference [between 1999 and 2013].
>
> (Revenue and Disaster Management Department, interview held on 24 July 2014, Revenue and Disaster Management, Bhubaneswar).

It has previously been argued that decision making, which is a crucial component of coordination in uncertain situations, can be hindered by poor planning, a lack of adaptive leadership and a culture of complacency (Chap. 2). This is elaborated in the context of Odisha by making three points. First, the reasons for poor planning in Odisha can be understood in light of the national landscape at that time. In 1999, India (let alone Odisha) did not have a disaster management policy. Also, the disaster management activities were fully within the jurisdiction of the state government.

After independence, the Government of Odisha inherited the colonial Famine Relief Code which they renamed the Scarcity Relief Manual to cover marked deteriorations in the agricultural season and crop failures caused by floods and rainfall (Ray-Bennett 2009). In 1980, after 33 years of independence, the Orissa Famine Codes of 1913 and 1933 were amended and the Orissa Relief Code (also known as the Relief Code) was conceived. The Relief Code was amended again in 1996 (GoO 1996). The Relief Code specifies how administrators should identify crisis conditions, how they should respond, and when they should do so during a natural calamity, as mentioned earlier (Currie 2000; GoO 1996; Ray-Bennett 2009). In the event of the Super-Cyclone, the Relief Code was instigated for the first time after independence (Ray-Bennett 2009; Samal et al. 2003). Since the disaster management system envisaged by the Relief Code was reactive in nature, it failed completely to deal with the complexity of the Super-Cyclone. There were no coordination mechanisms in place between the government departments prior to the Super-Cyclone. Also, the communication systems (both in terms of generating an effective early warning and disseminating) were under-developed. Hence, the failure of the coordination system was described as '*lack of plan and planning*' by the District Emergency Officer of Ganjam (see above) and '*no coordination*' whatsoever by the Director of the Indian Meteorology Department. This lack of coordination was hindered further because '*there was no authority to monitor relief and rescue*' operations from Bhubaneswar (District Emergency Officer, Ganjam). Lack of coordination was also acknowledged as a major failure during the Super-Cyclone by the Deputy Relief Commissioner of the Special Relief Organisation (22 July 2014, Bhubaneswar).

Second, this lack of co-ordination raises the issue of leadership—a critical component of decision making (Senge 1990). Critics blamed the then Chief Minister of the state, Mr. Giridhar Gamang for his weak leadership. He was unable to rise to the situation. The consequence of this weak leadership was such that Mr. Gamang and his Indian National Congress party was ousted from Odisha on 6 December 1999 by Biju Janata Dal—a regional party headed by Mr. Naveen Patnaik (Ray-Bennett 2009). Since then, Biju Janata Dal has won five successive elections achieving a record as the longest serving government in the history of Odisha (The New Indian Express 2015). Mr. Gamang remains as the shortest serving Chief Minister of Odisha. Some discussion on the leadership attributes of Mr. Patnaik in the case of Phailin is resumed in the next chapter.

Third, the coordination suffered further due to a culture of complacency which was rife in 1999. This was manifested in the form of unpreparedness, a reactive response system, inadequate measures for evacuation and a lack of imagination (conservative mind-set) amongst the district level responders. The District Emergency Officer of Jagatsinghpur and the Deputy General Manager of Odisha State Disaster Management Authority are quoted, in this regard:

> At that time we didn't have sufficient infrastructure, no warning system and we didn't think *ki* (that) such a […] disaster […] could hit Odisha coast because we didn't have an experience of this kind […]. Because of these reasons we had lost thousands of lives […] lost huge crores [hundred thousands] that ultimately had a bitter impact on the economic backbone of the country, [and of the] state.
>
> (District Emergency Officer of Jagatsinghpur, interview held on 29 July 2014, Swosti Hotel Lounge, Bhubaneswar).
>
> Prior to 1999 there was a disaster management and disaster preparedness but it was not ready. […] the system was pro reactive. Reactive means something happens then we react. So 1999 was the benchmark: why people died? There are two things – Odisha's disaster preparedness was prepared for a cyclone of much lesser magnitude. […] The number two, people had developed a complacency over a period of time.
>
> (Deputy General Manager of the Odisha State Disaster Management Authority, interview held on 23 July 2014, Rajiv Bhawan, Bhubaneswar).

A culture of complacency was also rife amongst the at-risk population by not heeding the early warnings due to a fatalistic mind-set that hindered evacuation (Samal et al. 2003). The evacuation process was further hindered due to a lack of shelters. In 1999, there were only 75 cyclone shelters on the entire coastline of 485 km (Harriman 2013; Ray-Bennett 2009). These shelters were built by the Red Cross Society and have saved thousands of lives. The culture of complacency was fuelled further due to a 'lack of experiencing' (Mingers 1980; Weick 1995) a devastating cyclone prior to 1999 (Samal et al. 2003). In this light, the Deputy Relief Commissioner of the Special Relief Organisation and the Deputy General Manager of Odisha State Disaster Management Authority are quoted below:

> Actually in 1999, so much of death was due to basically a want of shelter. Then number two, is people were not aware that such a huge calamity could come. So, they ignored the warnings of the authority and of the IMD (Indian Meteorology Department) […] because they have not faced such a calamity ever. That was new to them.
>
> (Deputy Relief Commissioner of the Special Relief Organisation, interview held on 22 July 2014, Rajiv Bhawan, Bhubaneswar).
>
> People had developed complacency over a period of time, so they ignored the warnings. When the administration [means communication] system gave the warning – a very big cyclone is coming, everyone was complacent – 'okay, let's see, we have lived with disasters, what happens'- but when the *hawa* (wind) blew 240–250 km/hr, they did not have an option but to die, that's it!
>
> (Deputy General Manager of the Odisha State Disaster Management Authority, interview held on 23 July 2014, Rajiv Bhawan, Bhubaneswar).

Thus, neither the responding actors nor the at-risk population imagined that a hazard of low-probability but of such great impact could affect the coast of Odisha. All of these factors caused the disjointed and unprepared disaster management systems to collapse.

3.3.2 Conservative World Views

The problems of coordination and communication are tightly coupled with world views. From the narratives above, it is evident the world views that impeded the disaster management system were not only due to a culture of complacency and reactive disaster management but also a lack of an 'overall objective' to reduce human deaths in the Super-Cyclone. The Chief Minister Mr. Giridhar Gamang, as a leader of the state and his bureaucrats, failed to generate an *objective* for the disaster management system. The consequence of this was more than 10,000 human deaths, of which 80% were in the district of Jagatsinghpur.

At an international level, in 1999 the proposal for proactive disaster management was in its early stages. This was also the case for the gender sensitive disaster response. It was only in the mid-term evaluation of the UN's International Decade for Disaster Reduction (IDNDR) (discussed in Chap. 1) in 1994 in Yokohama in Japan that the international community first acknowledged that women and men experience disasters differently (Enarson 1998; Ray-Bennett 2009).

Poor coordination, inefficient communication and an environment that lacked a culture of disaster preparedness both at home and abroad, became a seam at which the reactive disaster management system failed during the Super-Cyclone to save lives. A compilation of these views is provided in Table 3.5.

Table 3.5 Systems failure during the Super-Cyclone in 1999 (Compiled by author)

	Coordination	Communication	World views
Super-Cyclone 1999	No coordination between the government departments	Only via telephone	Complacent and conservative
	Core information was available only 48 h ago	Early warning system was under-developed	No goal was designed to reduce deaths for the disaster management system
	Odisha Relief Code was activated after the Super-Cyclone. Therefore, the response system was reactionary	The communication system to disseminate information and warnings was under-developed	Lack of political leadership
	No authority to monitor rescue efforts		A fatalistic mind-set that hindered evacuation
	Weak leadership		
	Lack of plan or planning		

References

Behera A, Sarkar HS (2003) Rising to the occasion: civil society response to the Orissa super cyclone, 1999. Orissa State Disaster Mitigation Authority Publication, Bhubaneswar

Bisoi KC (2013) Effective social mobilization for management of cyclone Phailin. The Response 2013(XIII):14–16

Census of India (2011) Orissa population census data 2011. Census of India. http://www.census2011.co.in/census/state/orissa.html. Accessed 5 Jan 2016

Comfort LK, Ko K, Zagoreck A (2004) Coordination in rapidly evolving disaster response systems: the role of information. Am Behav Sci 48(3):295–313. doi:10.1177/0002764204268987

Currie B (2000) The politics of hunger in India: a study of democracy, governance and Kalahandi's poverty. Macmillan Press Ltd, London

Enarson E (1998) Through women's eyes: a gender research agenda for disaster social science. Disasters 22(2):157–173. doi:10.1111/1467-7717.00083

GoO (1996) The Orissa relief code. Revenue Department, Bhubaneswar

GoO (2002a) Managing disasters in Orissa: background, challenges and perspectives. Orissa State Disaster Mitigation Authority, Bhubaneswar

GoO (2002b) Orissa state development report. Nabo Krishna Centre for Development Studies, Bhubaneswar

GoO (2005) State disaster management policy: resolution 4 Mar 2005. Revenue Department, Bhubaneswar

GoO (2008) District disaster management plan 2008–09: Cuttack. Emergency Operation Center (EOC) Collectorate Office, Cuttack

GoO (2013) District disaster management plan 2013–14: Jagatsinghpur District. District Emergency Operation Center, Jagatsinghpur

GoO (2014b) Ganjam disaster district management plan 2014–15, Chatrapur. District Emergency Operation Center, Chatrapur

Haeseler S (2013) Super cyclone Phailin across India in October 2013. Deutscher Wetterdienst (DWD), 18 Oct

Harriman L (2013) Cyclone Phyline in India: early warning and timely actions saves lives. UNEP Global Environmental Alert Services (GEAS). www.unep.org/geas. Accessed 20 May 2015

Mingers JC (1980) Towards an appropriate social theory for applied systems thinking: critical theory and soft systems methodology. J Appl Syst Anal 7:41–49

Mishra BN (2013) Tryst with Phylin: the deadliest cyclone in 2013. The Response XIII (Oct 2013):5–7

Ray-Bennett NS (2009) Caste, class and gender in multiple disasters: the experiences of women-headed households in an Oriya village. VDM Verlag, Saarbrucken

Samal KC (2003) Coping strategy of weaker sections (dalits and women) in post super-cyclone Orissa (case study of Ersamma block). Nabo Krishna Centre for Development Studies Publication, Bhubaneswar

Samal KC, Meher S, Panigrahi N (2003) Beyond relief disaster mitigation, livelihood rehabilitation and the post-cyclone recovery in Orissa: village level studies in three most cyclone affected districts in Orissa. Nabo Krishna Centre for Development Studies Publication, Bhubaneswar

Senge P (1990) The leader's new work: building learning organizations. Sloan Manag Rev 32 (1990):7–23

Swain M, Swain M, Sahoo RH (2006) Disaster and gender. Serials Publications, New Delhi

The New Indian Express (2015) Naveen Patnaik seeks self-appraisal of Ministers. The New Indian Express, (Bhubaneswar), 20 May

Weick KE (1995) Sensemaking in organizations. London: Sage

Chapter 4
Cycolne Phailin in 2013

Abstract This chapter presents the gender disaggregated death data from the four districts which were severely hit by the Cyclone Phailin in 2013. As with Chap. 3, it also provides the narratives of seven elite respondents which include the Director of the Indian Meteorology Department, Regional Director of UNDP, Manager of Odisha State Disaster Management Authority, Deputy Relief Commissioner of the Special Relief Organisation, and three Emergency Officers from Ganjam, Jagatsinghpur and Puri districts. Systems failure provides a novel perspective to analyse the reasons why there were fewer deaths in 2013.

Keywords Cyclone Phailin · Coordination failure · Communication failure · Deaths · Event violence · Gender disaggregated data · Goal · Justice · Leadership · Target groups · Worldviews · 'Zero casualty'

There were 86 deaths as a result of Cyclone Phailin. These deaths occurred in the four sampled districts. Cuttack had 19 deaths (10 males and 9 females); Ganjam had 36 deaths (22 males and 14 females); Jagatsinghpur had 18 deaths (9 males and 9 females) and Puri had 13 deaths (6 males and 7 females) (see Table 4.1). Although there were more than 10 deaths (unavoidable deaths), the consequences could have been far worse (Harriman 2013).

4.1 Coordination and Communication

Odisha was hit by cyclone Phailin on 12 October 2013 but the early warnings were disseminated on 8 October. This gave a lead time of 96 h to develop an effective disaster response system. According to the State Programme Officer of United Nations Development Programme:

> IMD (Indian Meteorology Department) played a crucial role. The information provided by IMD and Dr. XXX was really precise and really concrete because we always feared [what if] a lion comes (referring to a strong tropical cyclone). If you tell the people that a dog is

N.S. Ray-Bennett *Avoidable Deaths*, SpringerBriefs in Environmental Science,
https://doi.org/10.1007/978-3-319-66951-9_4

Table 4.1 Gender disaggregated mortality data in the cyclone Phailin 2013 (Compiled from the death records available at the District Emergency Offices in Puri, Jagatsinghpur, Ganjam and Cuttack 2014)

District	Male deaths	Female deaths	Total deaths
Cuttack	10 (53%)	9 (47%)	19 (100%)
Ganjam	22 (61%)	14 (39%)	36 (100%)
Jagatsinghpur	9 (50%)	9 (50%)	18 (100%)
Puri	6 (46%)	7 (54%)	13 (100%)

coming you put people in a difficult situation. But if you inform the people that a lion is coming, people will definitely move to safe places. This time this is what happened. They [referring to IMD] were very precise. I was in touch with them because the network was working. **Author: how did you manage to keep the network ongoing?** I was in touch with him the whole time. Every half an hour I rang him (referring to the Director of IMD) and in another half an hour Dr. XXX rang me up. **Author: what did you do with that information?** I was sharing this with our network and OSDMA (Odisha State Disaster Management Authority) [...] as part of NGO-GO Coordination.

(23 July 2014, United Nations Development Programme Office, Bhubaneswar).

The generation of this effective core information was facilitated by new technologies: high power computing systems[1] and the Meteo France International Synergie System. These super computers and the Meteo France International Synergie System enabled quick and accurate predictions after analysing weather data (Barik 2014).

After the core information was generated and verified by the Duty Officers and their associate networks in New Delhi, the Director of the Indian Meteorology Department in Bhubaneswar emailed and faxed this information to the Special Relief Commissioner at Special Relief Organisation. The Indian Meteorology Department also disseminated this core information to Odisha State Disaster Management Authority; United Nations Development Programme Office; the Chief Minister; media through direct interviews; and District Collectors, Block Development Officers, District Emergency Officers via mobile texts and email messages. However, the Indian Meteorology Department in Bhubaneswar did not target the at-risk community directly (Director of Indian Meteorology Department, 21 and 22 July 2014). Additionally, the core information that the Indian Meteorology Department generated was not an 'alert', but rather an 'early warning'—one that

[1]High power computing systems and the Meteo France International Synergie System were bought in 2009 and 2010 respectively. The author was able to see these super computers under the guidance of the Director. The Meteo France International system operates through three connected monitors. It provides real time wind, rain and air pressure forecasts of 48 h. In the case of power-cuts, or a failure of the internet, this system continues to work without disruption due to its connection with the INSAT Satellite. During Phailin, this highly developed technology enabled the Director and his Duty Officers to generate almost accurate early warnings (Director of Indian Meteorology Department, 21 July 2014; Barik 2014).

suggests that “a hazard is occurring or is imminent. The public should take action” (Herrmann 2009: 11). This is explained in the words of the Director of Indian Meteorology Department:

> I faxed as well as emailed messages to SRC (Special Relief Commissioner). […]SRC then transmitted this information to District Collectors who then targeted their respective districts, people in the low-lying areas and so on […]. For this there is usually a time gap of two hours. By that time our messages reached to all the District Collectors because we have their email ids and mobile numbers. So they reached directly to them. […] this message is generic only, not an alert. Any alert […] I am not the authority. SRC is. SRC should alert messages to District Collectors for relief, rescue, food aid or the like […]. We have a hotline between SRC and my office. We talk.
>
> (22 July 2014, Indian Meteorology Department Office, Bhubaneswar).

The Special Relief Commissioner at the Special Relief Organisation issues weather ‘alerts’. A weather ‘alert’ is “the highest level in the warning system, issued when a hazard is sure to occur” (Herrmann 2009: 12). Although weather ‘alerts’ are issued by the Special Relief Commissioner, the overall act of issuing an alert is a complex process because related responsibilities sit across various departments and amongst various decision makers. This was noted by the Director of the Indian Meteorology Department; ‘we *have a hotline between SRC (Special Relief Commissioner) and my office. We talk*’. Hotlines (point-to-point communication link) were unavailable in 1999. In 2013, during Phailin they facilitated inter-departmental communication.

Due to the ‘Disaster Management Act 2005’, the Odisha State Disaster Management Authority has been set up by the State Cabinet exclusively to assess a disaster situation and make appropriate decisions and strategies accordingly (GoO 2005; OSDMA 2011, 2014). This Committee is headed by the Chief Secretary under the leadership of the Chief Minister Mr. Naveen Patnaik. The role of the Special Relief Commissioner in this regard is to coordinate these meetings and make prompt decisions (including related to the issuing of an alert) in response matters (GoO 2005). This is evident below in the narrative of the Deputy Relief Commissioner of the Special Relief Organisation. In this context, the role of the Special Relief Organisation is important to kick-start the response system on time by activating the Disaster Management Plans and Emergency Operation Centres in the districts (GoO 2005).

> In the case of Phailin, we got official information on the evening of 7th. From 8th, there were constant coordination meetings and reviews at the level of Chief Secretary and the Chief Minister for the following four to five days. My department takes part [in these meetings], […] comply and implement. In these meetings, usually proceedings [or ordinary formalities] are not necessary. They give orders directly. **Author: This is very interesting!** Yes, if it is left for writing letters then it would take time to reach the concerned person […].
>
> Another thing that is happening now is while IMD (Indian Meteorology Department) gives us information, simultaneously they feed into media. […] This helps people, Collectors, BDOs (Block Development Officers) and Tahasildars to prepare. At village level, loudspeakers are also arranged to warn people. […] Media is very aggressive these days. Lots of competition which is a good thing. Media has been very helpful […] during Phailin. The

Special Relief Commissioner and our Chief Minister – were able to communicate their messages to the people through media […].

We also manage our emergency operation system round the clock by two persons. During emergency four people. I am also there during emergencies. This system activates the District Emergency Operation […] and suggests ensuring preparedness including food, boat arrangement […].

Fax and emails were unavailable in 1999. Telephone facility was there but only with the Emergency Officer in Cuttack […]. Mobile phones have helped a lot during Phailin. The Special Relief Commissioner asked me to collect some information from the Collectors. But by the time I have rang them up, I learnt that they have already spoken to the Chief Secretary or the Additional Secretary […] this is a new phenomenon. Ongoing dialogue […].

Author: Does this cause duplication and confusion? […] It is good. Nothing will be ignored or lose sight of. If something has been left out by the SRC (Special Relief Commissioner), may be it will be caught by the Chief Secretary. Perhaps this causes some duplication, but can be avoided […]. Some kind of redundancy is needed for disaster management […] [with regard to] communication channels. I should have four or five types, if something fails, the other will work. […] That means I should have many channels open, not a single channel to talk to people […]. What if it fails […] the entire system will fail.

(Deputy Relief Commissioner of the Special Relief Organisation, 22 July 2014, Rajiv Bhawan, Bhubaneswar).

The Deputy Relief Commissioner's excerpt indicates that the standard operating procedures were dispensed with in the disaster climate for Phailin because of time constraints. Such an approach was taken because timing and critical decisions are crucial in a disaster climate if lives are to be saved. Hence, formalities related to bureaucracy were relaxed. This is also noted by the responders in the districts later in this chapter. This practice demonstrates a degree of agility—something that is pivotal to developing an effective response system to minimise fatalities in a disaster climate.

Second, it is also noted that an official announcement of a weather alert kick-starts the Emergency Operation Centres in the districts. This established a new administrative network, one that did not exist during the Super-Cyclone.

Third, the narrative suggests that although multiple communication channels can confuse and duplicate information, responders are encouraged to do so in order to promote dialogue for an effective response system.

Lastly, the Deputy Relief Commissioner's quote: "*Nothing will be ignored or lose sight of. If something has been left out by SRC (Special Relief Commissioner), may be it will caught by the Chief Secretary*" is compelling. It confirms the existence of a new socio-technical disaster management system in Odisha that was not present in 1999 (GoO 2005). Actors in this new system are continuously conversing and are encouraged to do so in order to align with the state, district, block and village levels. Mobile phones, hotlines, e-mails, landlines, HAM radios, V-Sat—all these communication tools were major enablers for this change. The role of social media such as Facebook or twitter, however, were not emphasised by the respondents.

Furthermore, actors were aware of synergies and interdependencies and they demonstrated this by supporting and complementing each other's weaknesses and strengths (Lalonde 2011) during Phailin. This dynamic interdependencies combined with an agile bureaucracy were vital to minimising the impact of Phailin on people.

4.2 Coordination and Communication at the Districts Level

As specified by the Deputy Relief Commissioner, the district level authorities received the early warning information on 8 October. In this regard, the former District Emergency Officer of Ganjam and the District Emergency Officer of Jagatsinghpur is quoted:

> […] on 8th October we were intimated by IMD (Indian Meteorology Department), one severe cyclone may hit Ganjam coast. […] they did not confirm […] an 'alert'. On that day we called an emergency meeting with all the line departments […] [including] the irrigation department, […] health department, the CDMO (Chief District Medical Officer). For control of law and order situation I intimated SP (Superintendent of Police) […], Public Works Department […] Civil Supplies Corporation (for steady supply of essentials such as food in the district) […]. Also we intimated the telephone tower people to be ready with an alternative use of […] communication during emergency […]. This is called the District Disaster Management Committee meeting.
>
> **Author: how long did it take to arrange this meeting?** Only four hours. […] Within four hours we called all the line department officers […] to shoulder responsibilities. […] All came. CBDO (Community Based Disaster Organisation), SDO (Sub District Officer), Chief District Veterinary Officer, District Deputy Director of Agriculture […]. So, we all were present in the meeting. […] we have done, you know, systematic way from the day-one.
>
> By interacting with our government officers, on that day, we have given them the instructions to prepare their own plan – a micro plan. […]. On 9th we have received the micro-plans […] and intimated our BDOs (Block Development Officers) to prepare evacuation plan which are the villages 0–5 kms from the seashore […]. We have asked people who are in *kutcha* houses (mud houses) must shift to shelters […] [but] leave all those who are in *pukka* buildings (concrete) […]. All BDOs had prepared micro-list – […] also a list of rural officers who are responsible to shift people. They […] have phone numbers, list of buildings and so on. […] we instructed BDOs, Sub-Collectors, one person from NGO, local volunteers and Sarpanch[2] you know to visit villages on 9th […] and make a mike announcement by 10th. We ensured all people [received the warning prior to the landfall of Phailin].
>
> On 11th, you know by 2.30 pm almost 3–4 lakhs (or 300–400,000) people were evacuated. We have that list. They were shifted by bus, truck all conveyance […] for free. […] we have maintained records […] of all persons […].

---

[2]*Sarpanch* is an elected head of a village.

> […] the government also blindly approved all proposals from this district […] in utilising all vehicles […] resources […] for those who are residing in 0 to 5 kms […]. It was a hectic task […] in carrying out all our instructions very effectively. […].
>
> There are 15 cyclone shelters […]. Actually […] 15 cyclone shelters are not enough, you know. You cannot accommodate 1 lakh (100,000) people in 15 cyclone shelters. **Author: what did you do?** We selected all *pukka* (concrete) Sarva Sikshya Abhijan buildings, *pukka* buildings which are located 5 kms away from the sea. We located all immediate schools, college, office buildings, private engineering colleges […]. We also provided safe, you know toilets cistern, also provided some kitchen facilities, free kitchen […].
>
> […] some people stayed [in the shelters] for five more days […] so we also provided aftermath relief, food packages, dry rations […] of 50 kg rice, condiments and other purchases, five hundred rupees […] until [the return of] normality.
>
> (Former District Emergency Officer of Ganjam, 24 July 2014, currently deployed at the Revenue and Disaster Management Department, Bhubaneswar).

According to the District Emergency Officer of Jagatsinghpur:

> SRC (Special Relief Commissioner) and the [other] highest quarter – […] emphasises 'District Collector don't sit alone, bring […] all departments and […] plan accordingly' - this is […] the district level Disaster Plan. It is a coordinated effort that is in tangible form. **Author:[…] so it is happening?** Yeah it is happening. It is very much happening. Otherwise it is very difficult for a one man sitting at the district [to mitigate the effect of Phailin] – it is after all inter-departmental efforts […].
>
> In villages, if disaster happens, there are people in charge […]. So, this is an additional duty that they are assigned [as part of the Disaster Plan] […]. This is the outcome of inter-departmental coordination and District Disaster Management Committees. […]
>
> Besides you are adding other important dimensions like the involvement of NGOs and the village communities, the elderly persons […] inviting them to our meetings […]. We are noting them down and we are implementing their every important suggestion because experience makes a man perfect. So they are having perfect experience, we are gaining from their knowledge.
>
> (29 July 2014, Swosty Hotel Lounge, Bhubaneswar).

The availability of this core information prior to Phailin was vital in guiding the primary responders' actions and facilitating their sense making processes in an uncertain situation (Ayan et al. 2015; Harriman 2013). The outcome of this was that the responders were able to evacuate 1.2 million people from 18 coastal districts (Mishra 2013; Harriman 2013). In the sampled districts about 114,306 people were evacuated from Cuttack, 180,000 from Ganjam, 101,457 from Jagatsinghpur and 102,000 from Puri (Nayak 2013). This evacuation operation is considered as one of the largest operations in India in 23 years (Harriman 2013). An operation of this scale was only possible because of the coordination between actors, the availability of core information, the delegation of authority to the primary responders, effective evacuation planning, flexible bureaucracy, willingness of the affected community, and the sheer dedication and commitment of the responders to save lives.

The narratives above also suggest that the alignment of the communication sub-system was possible due to a much improved administrative network and knowledge network established in the 'inter-disaster period[3]' (period between 1999 and 2013) in Odisha in particular and India in large (Ayan et al. 2015; Srivastava 2009). This network of actors and organisations provided much needed direction and power to the responders in the districts and villages in a disaster climate. Most importantly, this system also benefited from the involvement of the NGO sector, media, mobile, radio, at-risk people and volunteers.

Mobile phones and the radio, in particular played a crucial role to warn the general public. Mobile phones provided up-to-date information about the severe weather conditions. In 1999, mobile phones were a rarity in Odisha. Currently there are more than 20 million users, while the state-run Bharat Sanchar Nigam Limited (BSNL) alone has three million subscribers. Bharat Sanchar Nigam Limited texted early warnings to 1.5 million people, mostly in the coastal belt, on 11 and 12 of October (IANS 2013). The state-run broadcaster All India Radio (AIR), on the other hand, reached more than 80% of the rural population before Phailin's landfall (IANS 2013). These collaborative efforts were appreciated and acknowledged by the government actors. These efforts were also conducive to generating adaptive and collective leadership amongst the responders who carried out relief, response and evacuation operations. The outcome of this network space of actors and organisations was a radical reduction in human deaths in Phailin. Odisha is now dubbed as a model for India, as well as to the whole world (Ayan et al. 2015). The World Bank praised Odisha by saying:

> Successfully evacuating a million people is not a small task. This cannot be merely achieved by kicking the entire state machinery into top gear for 3 to 4 days following a cyclone warning. This has taken years of planning, construction of disaster risk mitigation infrastructure, setting up of evacuation protocols, identification of potential safe buildings and most importantly working with communities and local organizations in setting up volunteer teams who all knew exactly what needed to be done (Nayak 2013: 8).

The Director of UNISDR, Margareta Wahlström also praised Odisha by saying:

> Odisha's handling of the very severe cyclone will be a landmark success story in disaster management. We are very impressed. We have plans to use it as a model for other cities and countries to follow as part of our global efforts on disaster risk reduction (Nayak 2013: 12).

4.3 Enlightened World Views

By 2013, Odisha had come a long way. The State Programme Officer of United Nations Development Programme Office pointed out that:

> Post 1999 it was not only the UN Organisations, but also the national and non-governmental organisations – both local and international- all parts came together to

[3]The District Emergency Officer of Jagatsinghpur described the period between 1999 and 2013 as an '*inter-disaster period*'.

> respond because there was a large number of human casualties in the Super-Cyclone […]. Since then, there has been a significant change in the understanding of the disasters. An average Odiya, even in the rural areas are much aware as how to respond or prepare in the early warning of disasters. […]. UNDP took a lead role in terms of facilitating community based disaster plans in almost close to 24,000 villages. **Author: interrupts, is this the Disaster Risk Management (DRM) programme?** Yes. DRM created a shift in the mind-set of the community […]. Working with the communities and then that reflection process helped building their capacities.
>
> […] before Phailin, people used to come and ask me 'what about the DRM programme?' I gave this example 'you see when a dish is cooked, any capacity building programme is just playing a role of adding salt for taste. After the dish is cooked if you ask me 'where is the salt?' I cannot show you. You only taste it when you eat it'. So any capacity building programme aimed at bringing any intrinsic changes, one cannot show you any regular or normal features. **Author: is this a salt phenomenon?** Yes, you cannot show it because it has gone to the veins. It has gone into the DNA […].
>
> (Interview held on 24 July 2014, United Nations Development Programme Office, Bhubaneswar).

The metaphor of a 'salt phenomenon', applied to 'capacity building' processes, is a simple and yet powerful analogy in the context of Odisha. Building the capacity of responders, organisations and the at-risk community involves slow and process oriented activities. They take time, significant effort and continuous reflection and action. The State Programme Officer mentioned above that the United Nations Development Programme Office, as an international player, "*played a significant role to keep this issue alive*". United Nations Development Programme Office and the Government of India together launched the Disaster Risk Management (DRM) programme between 2002 and 2012 and it helped to build the capacity of the responders and the at-risk community.

The salt phenomenon also counterbalanced a culture of blame. All the relevant actors genuinely began to improve the disaster management system by learning from their mistakes rather than blaming a failed system and a failed leadership during the Super-Cyclone. The Deputy General Manager of Odisha State Disaster Management Authority is quoted in this instance:

> We committed mistakes in 1999. Humans commit mistakes. So, this time we wanted to learn because we experienced very horrific situation of losing 10,000 lives […]. People who were alive, they could not be provided with medicine, water and food in time because our system of restoration or coming back to normal was overworked.
>
> We did not have a good disaster management system in place in 1999. We did not have inflatable rubber boats to rescue people from flood, now we have more than 300. But that is not important – important is we have learned and trained our people to respond in their own ways […].
>
> (Interview held on 23 July 2014, Rajiv Bhawan, Bhubaneswar).

In this inter-disaster period, the capacity building process has crept into the DNA of Odisha's social system and the socio-technical disaster management system. The effect of this was also recognised at a global level in 2011 when Odisha's capital, Bhubaneswar city won the UNISDR's Sasakawa Award in Switzerland for a 'Role

Model City in Community Preparedness and Making the City Disaster Resilient' (OSDMA 2011: 23). Unsurprisingly, the effects were remarkably noticeable during Phailin when a poor state like Odisha reduced human deaths from 10,086 in 1999 to 86 in 2013.

The machinery of government was able to think ahead and set up a collateral organisation, such as Odisha State Disaster Management Authority in the immediate aftermath of the Super-Cyclone—dedicated solely to coordinate and communicate disaster management and mitigation issues in the state. Gujarat, Andhra Pradesh and Uttaranchal were the other Indian states who adopted this model by setting up collateral organisations for disaster management (Srivastava 2009). Odisha was also the first state to pass a bill for a state disaster management act, one which was later adopted by the Government of India as a national 'Disaster Management Act 2005' (Deputy Relief Commissioner, July 2014; Ray-Bennett 2009).

Odisha was also a step ahead in creating Odisha Rapid Action Force (ORAF) to support the district disaster response system. Odisha Rapid Action Force assists the district administration, acts as a first line of support, moves prior to a disaster climate, locates equipment in strategic places, conducts search, rescue, evacuation and relief, and reduces dependencies on army/paramilitary forces and also minimize expenditure and time (GoO 2001). Odisha Rapid Action Force was launched on 7 June 2001. During the time of fieldwork, there were 10 Odisha Rapid Action Force units and there were discussions to develop another 20 of such Force Units (Deputy Relief Commissioner, July 2014). This should allow a presence of at least one Odisha Rapid Action Force Unit in all the 30 districts of Odisha. Each Odisha Rapid Action Force contains 50 expert personnel and they played a crucial role in evacuating more than 1.2 million people prior to the landfall of Phailin from 18 districts (Mishra 2013; Deputy Relief Commissioner, July 2014). The Government of India has also adopted this model by forming a National Disaster Response Force (NDRF) (Deputy Relief Commissioner, July 2014). Both Odisha Rapid Action Force and National Disaster Response Force together played a crucial role in evacuating people during Phailin (Ayan et al. 2015).

The myriad of capacity building programmes for the responders, volunteers and for the at-risk communities by Odisha State Disaster Management Authority, United Nations Development Programme Office, Government of India, INGOs and NGOs—jointly—led to a 'culture of disaster preparedness' (OSDMA 2011, 2012), one in which progressive world views are nourished. These world views include gender sensitivity, building human capacities, agile bureaucracy in a disaster climate, willingness to learn from mistakes, support for the use of technology and a strong commitment to reduce human losses—amongst many others. This mind-set stood in stark contrast to the conservative mind-sets observed in 1999. These mind-sets were also the outcome of reflection, action and interdependencies within and with the outside world. This is explained in the words of the State Programme Officer of United Nations Development Programme Office again:

> We have learnt a lot from South-South collaboration. For instance, Bangladesh experience. […] learning has evolved. Post 1999 there were large numbers of educated people. They

came and started working in Odisha in disaster response programmes. This is a big difference. When educated person, volunteers to work with the communities, he or she facilitates l…] reflection process. That really helps in facilitating reflection by the communities and also helps in building the knowledge and add to the body of knowledge. This was hugely contributed by these people who volunteered to work post 1999.

It also created […] opportunities for them to go and work with other international organisations. So if you go to any international organisations, primarily NGOs, if you see who is leading or managing the disaster programmes, you will find [volunteers from Odisha]. We have learnt from each other […].

(23 July 2014, United Nations Development Programme Office, Bhubaneswar).

4.4 Leadership for Avoidable Deaths

Leadership is central to promoting a culture of disaster preparedness. Mr. Naveen Patnaik, the Chief Minister of Odisha provided much needed leadership in the aftermath of the Super-Cyclone. In addition to observing a state level 'Disaster Preparedness Day on 29 October', he also concentrated much of his effort in building the infrastructure of his state—one that is essential to supporting a disaster response system. This is thanks to the funds available from the World Bank and the central government in order to build roads, bridges, concrete houses, multi-purpose cyclone shelters, schools and the like (State Programme Officer of United Nations Development Programme Office, 23 July, 2014; Ayan et al. 2015; Ray-Bennett 2009). Good road conditions, as well as their connectivity with the cyclone shelters facilitated the evacuation process during Phailin (State Programme Officer, United Nations Development Programme Office, 23 July 2014; Deputy Relief Commissioner, 22 July 2014).

He also exhibited the traits of a strategic leader by developing a 'shared vision' (Senge 1990) or 'a goal' (Jenkins 1969) of 'saving precious lives' for all actors involved in mitigating the effect of Phailin. This remarkable success in reducing deaths in Phailin heightened Mr. Patnaik's popularity amongst his electorates. The Telegraph called this a 'Naveen Factor' (Mudur et al. 2013).

Mr. Patnaik is also a charismatic leader. At the age of 68, he has never been married. Odiya people believe a man with no family connections has no greed and therefore, has no desire for corruption. Hence, a never-married man is regarded to be an ideal statesman who can dedicate his life solely to the greater good of human-kind. This belief-system is deeply rooted in the Indian philosophy of *brahmacharya* (sexual self-restraint) and *sannyasa* (renunciation of material world and greed) (Prime 1992). These are highly desirable qualities for a leader in Indian politics. The 11th Prime Minister of India Mr. Atal Vihari Bajpai, the current Prime Minister Mr. Narendro Modi and the Chief Minister of West Bengal Miss Mamata Banerjee (to mention a few) have all won elections based on these qualities. Mr. Patnaik's immersion in politics and his dedication to improving the wellbeing of his state is admired even by the bureaucrats that the author met. The participants often referred to Mr. Patnaik as 'our Chief Minister', 'a good Chief Minister', 'a very dedicated Chief Minister who arranges meetings regularly' or the 'people's Chief Minister'. As a charismatic leader he has won the hearts and minds of Odiya people.

Nevertheless, it can also be argued that his leadership tactics (both strategic and charismatic) are politically motivated—based on winning votes and remaining in power (Mudur et al. 2013). Mr. Giridhar Gamang's downfall in the Super-Cyclone led to the rise of Mr. Naveen Patnaik. This is noted by Mudur et al. (2013):

> Naveen knew complacency could cost him dear with elections barely six months away. The rapid response over the weekend contrasts with the events in 1999. So numbing was the effect of the super cyclone that nothing moved in the state for nearly 12 hours after landfall, not even relief trucks parked outside Congress chief minister Giridhar Gamang's residence. When they did leave for their destinations, they were looted on the way.

Since 1999 Mr. Patnaik has won five successive elections. He is the longest serving Chief Minister in the history of Odisha. Hence, setting a goal 'to save lives' in Phailin for the disaster management system may also be considered as a political act —one that will secure him and his cabinet another tenure of leadership in Odisha. In the context of this research, however, this politically motivated act was beneficial because this was translated into a rational goal for the entire disaster management system by the Special Relief Organisation. This is explained in the words of the Deputy Relief Commissioner of the Special Relief Organisation:

> If you take the case of Phailin, we got official information on 7th evening. From 8th there were constant coordination meetings and reviews at the level of Chief Secretary and the Chief Minister for the next 4 to 5 days […]. **Our Chief Minister […] suggested life is precious and we must save them […]** [author's emphasis].
>
> (22 July 2014, Rajiv Bhawan, Bhubaneswar).

To implement the Chief Minister's order, the Special Relief Commissioner conveyed the goal of a 'zero casualty' to all the responders in the district, block and village level which became a mission statement for the state machineries (OSDMA 2014). The verbatim copy of this mission statement is produced in the box below.

Zero casualty: The mission of the state government

Considering the staggering number of deaths due to the Super Cyclone in 1999, preventing casualties to human lives became the primary objective this time. Zero casualties became the war cry for the state government and preparedness planning revolved round this. Chief Minister took lead in cyclone preparedness planning by

- Interacting with district administration through video conference and guiding them;
- Writing to The Union Defence Minister and National Disaster Response Force (NDRF) requesting for the deployment of personnel to carry out rescue and relief operations;
- Monitoring the preparedness from time to time by presiding over review meetings between 09.10.2013 and 12.10.2013; and
- Appealing on air and through newspapers to the people of Odisha not to panic but to cooperate with district administrations in evacuation operation.

(OSDMA 2014: 57)

The effect of this goal was that a flexible bureaucracy was enabled, one that did not follow everyday standard operating procedures to seek human and financial resources. The Special Relief Commissioner also removed structural barriers in order to optimise the efforts and actions required to reduce human deaths 'at any cost' during Phailin. Three District Emergency Officers are quoted in this context because they translated this goal into action:

> The Collector of Ganjam [...] SRC (Special Relief Commissioner) and the Chief Secretary - together - they took one decision [decision to save lives]. The Chief Minister instructed that there should not be deaths, our motto should be [...] zero casualty. So accordingly we carried out instructions.
>
> In some places, we convinced them (referring to village people), and in some other places we forced them [...] because [...] our main motto was to save life. We cannot save properties of all. We wanted to save lives at any cost, so we used also force which were also appreciated later on by those people. **Author: What kind of force?** Just warning. On 10th evening people started [to evacuate] when they saw [...] dark clouds, breeze [...].
>
> (Former District Emergency Officer of Ganjam, 24 July 2014, currently at the Revenue and Disaster Management Department, Bhubaneswar).

According to the District Emergency Officer of Puri:

> [...] the instruction of the Collector was that, do anything but deaths should not be there, deaths should not occur. Do anything – [...] you can incur any type of or any expenditure but make sure that there should not be any death. [...]. All BDOs (Block Development Officers) and all Tahasildars, all field level Officers took every sort of steps to check deaths. [...].
>
> We had only one objective – [...] there should not be any casualty. The message came from the government; it is from the CM (Chief Minister). Care should be taken for zero casualty.
>
> We cannot stop it also, cannot stop it. We were having four casualties at that time. Only four. One was 78-80 years. Old man. [...] you know elderly people, they have affinity to stay in their houses [...].
>
> (31 July 2014, District Emergency Cell, Puri).

According to the District Emergency Officer of Jagatsinghpur:

> Our strategy was to ensure no casualty, no death. At least, at least, that was the motto. Keeping this motto in our mind we have managed with our, of course, very low grade infrastructure. We don't have sufficient upgraded technology. Whatever indigenous technology we have at the district level, we have managed.
>
> [...] after we have received the information from the Special Relief Commissioner and IMD (Indian Meteorology Department) that our district is going to be hit by a cyclonic storm [...] the first thing I did was organised a meeting within two hours [...] because everybody is having mobile [phones] these days. There is a big hall in our Collectorate and the Collector presided over the meeting and gave some instructions [...] to achieve zero casualty. There are strategies for this which we have circulated [...] i.e. the District level Disaster Management Plan [...]. It is a voluminous book. We also have micro plans [...] because few pockets, few areas [...] are more vulnerable [...] we have learned from experience [...] not intuition. [...].

> Involvement of community is [equally] first and foremost. Because of lack of community involvement we have failed [...] miserably during the Super-Cyclone. [...] They (referring to community) had a mind-set that it was the government's duty to supply medicine, water [...]. But now they think only two to three persons are taking care of us, what can they do - better we should help them. This is a new mind-set [...] in Odisha.
>
> Another important thing that we have done is the role of NGOs and voluntary agencies. They have helped like anything. Without any financial or any kind of interest, I have seen lots of NGOs readily [...] rescuing cows, buffalos, humans [...] that is the reason we could aim for zero casualty. [...] up to 50,000 people could have died [...] instead 12 persons died. This is a miraculous change we achieved during this inter-disaster period.
>
> The efforts from the public, the community, [...] bureaucracy, [...] government and the role of NGOs in between – makes a vibrant society and that ultimately had an impact on our mind-set. Our mind-sets are changed now. We are feeling comfortable *ki* [that) how big the disaster is, let the disaster come – we are ready.
>
> (29 July 2014, Swosti Hotel Lounge, Bhubaneswar).

The above extracts assert that district response systems were geared towards achieving 'zero casualty' at any cost. At the same time achieving 'zero casualty' or only 'unavoidable deaths' is an ideal goal in general and for resource constrained government organisations in particular. This was reflected upon by the District Emergency Officer of Puri when he mentioned "*We cannot stop it also, cannot stop it. We were having four casualties at that time. Only four*".

The extract of the District Emergency Officer of Jagatsinghpur also suggests that a target of zero casualty cannot be achieved without involving the affected community, media and the NGO sector. As a result, their efforts were vital to averting the risks of Phailin.

It is suggested that the inter-dependencies of actors, the culture of disaster preparedness and the clear goal set by the Chief Minister to reduce deaths led to an enlightened perspective in Odisha—one that is resilient and reflective. This was noticeable in the quote of the District Emergency Officer of Jagatsinghpur:

> Our mind-sets are changed now. We are feeling comfortable *ki* [that] how big the disaster is, let the disaster come – we are ready.
>
> (29 July 2014, Swosti Hotel Lounge, Bhubaneswar).

4.5 Target Groups

Although the current disaster management system in Odisha exhibited progressive, resilient and reflective traits, its response system is still based on a world view which is gender-neutral. However, gender-neutrality does not mean that the system is not gender sensitive. The response systems in the districts were gender sensitive but the early warnings and alerts generated by the secondary organisations, such as the Indian Meteorology Department and Special Relief Organisation, were gender neutral.

As mentioned earlier, the Indian Meteorology Department in Odisha does not generate core information tailored to the at-risk community. Their target groups include not only the Special Relief Organisation, United Nations Development Programme Office, District and Block Level Disaster Management Authorities, but also the fisheries, agriculture, railway, aviation and navigational advisories, defence and spacecraft launching services and expeditions related to the Antarctic and pilgrimage (Indian Meteorology Department/Odisha 2014; Director of Indian Meteorology Department, 24 July 2014).

Special Relief Organisation on the other hand, whilst they are the lead organisation to activate the district disaster response system by generating a weather alert, they also do not target the at-risk community directly. Instead their target groups are the District Collectors, Odisha State Disaster Management Authority, Block Development Officers, Tahasildars, and District Emergency Officers. This is explained in the words of the Deputy Relief Commissioner:

> **Author: do you customise your target group when issuing an alert or when you speak with the media?** Not really. Actually people are getting enough information from media. […] we advise people to keep radios on in the absence of electricity.
>
> We advise […] the District Emergency Operation to activate immediately – to ensure all preparedness arrangements are in place […] food, boat, and strategies etc**., Author: will the district authorities decide the target groups?** […] we used to keep track of them but it is difficult from here [Bhubaneswar] to do so. They have to implement this, who they want to target – a small village or villager. We have Collectors, below them Sub-Collectors, Block Development Officers, Tahasildars and also some other officers on the ground. […] basically they exercise their authorities.
>
> (22 July 2014, Rajiv Bhawan, Bhubaneswar).

This practise of generating early warning and weather alert is inconsistent with the Sendai Framework for Action which suggests to “develop, maintain and strengthen people-centred multi-hazard, multisectoral forecasting and early warning systems […]; tailor them to the needs of users, including social and cultural requirements, in particular gender […]” (UN 2015: 21).

Whilst interviewing the three District Emergency Officers, it was discovered that the response system in practice is gender sensitive. This is largely thanks to the myriad gender sensitive community based disaster programmes that were initiated by United Nations Development Programme Office, Action Aid and Odisha State Disaster Management Authority in the aftermath of the Super-Cyclone. This also led to a culture of gender awareness (OSDMA 2012; Ray-Bennett 2009, 2010; State Programme Officer of United Nations Development Programme Office, 23 July 2014). A gender sensitive response system is at the behest of the ground level responders, the District Collectors and the Emergency Officers. They can target vulnerable men or women specifically, but if they do not, the system will not penalise them.

Since the main objective of the disaster management system in Phailin was geared towards achieving ‘zero casualty at any cost’, it was noted that such a system was beneficial for the vulnerable groups. Women, children and elderly people were evacuated first, which is evident in the narratives of the two responders from Jagatisinghpur and Ganjam districts below:

> Life is precious. We wanted to save all lives. In Ganjam in particular we found that actually less women died in disasters. Actually women were the first who came out from their villages […]. They told 'we will go to shelter houses to save our lives'.
>
> (Former District Emergency Officer of Ganjam District, 24 July 2014, Revenue and Disaster Management Department, Bhubaneswar).
>
> This time what happened, we have taken care of pregnant women and girl children. We have instructed the field employees like the Tahasildar, BDOs (Block Development Officers), Gram Sevak (village worker) Asha Karmi on this […]. I told them categorically women and children must be our top priority and then old men and we should take help of the youth for protecting […] relief camps […].
>
> In the cyclone shelters […] we have initiated a separate room for […] women […]. Those are the simplest intervention taken by our Asha Karmi […].
>
> (District Emergency Officer of Jagatsinghpur, 29 July 2014, Swosti Lounge, Bhubaneswar).

The narrative of the District Emergency Officer of Jagatsinghpur in particular suggests that the grassroots responders were given strict instructions to evacuate women and children first. These instructions were facilitated by the decision taken by the Special Relief Commissioner to enforce the Clause 34(C) of the 'Disaster Management Act 2005' in the event of Phailin. This particular Clause gave power to the district and local level responders to evacuate people 'by force' if needed (Mishra 2013).

4.6 Justice for Event Violence

According to the State Programme Officer of United Nations Development Programme Office, the current disaster management practice is built upon a rights-based approach—one in which '*people's dignity are respected in terms relief and rehabilitation*'. This approach has created heightened awareness amongst the '*average Odiya person even in the rural areas with regard to response and preparedness*' (State Programme Officer, United Nations Development Programme Office, 23 July 2014) and also about their '*entitlements*' (Deputy Relief Commissioner, July 2014).

In the inter-disaster period, death compensation played a major role in promoting peoples' entitlements.[4] Although a deceased person is irreplaceable, compensation enables the state government to acknowledge the event violence that occurs during disasters. Although the idea of event violence presented in Chap. 1 is concentrated primarily around the failure of the disaster response system to protect lives, this was

[4]According to the Orissa Relief Code (GoO 1996), deceased family is given death compensations. These compensations are hazard specific.

Table 4.2 Rectifying systems failure (Produced by author)

Coordination
Core information was available >96 h ago
"excellent coordination" between State, District, Block Departments and Indian
Meteorology Department
Enhanced knowledge and administrative networks between the Government of India and Government of Odisha and the UN
• Synergies and inter-dependencies of responders and at-risk community • Coordinated effort to issue weather warning and alert on time • Effective micro and situational plans by the Category 1 responders • Effective enforcement of disaster management plans • Effective coordination between the State Emergency Operation Centre, the Control Room at the Special Relief Organisation and National Emergency Communication Network • Effective coordination of NGO-GO-Village Level Task forces through meetings • Effective involvement of the media, NGO sector, and the at-risk community • Effective meetings with the Disaster Response and Volunteer Team organised by the International Federation of Red Cross and Crescent Society • District Disaster Management Committees were important hub to direct and lead the block, village level officers, NGOs and volunteers for evacuation and relief • Excellent support from the Odisha Rapid Action Force and National Rapid Action Force in evacuating people
Communication
Administrative and knowledge network due to the Disaster Management Act 2005
Mobile, TV, radio, email, fax, hot line, deployable V-SaT terminals, satellite phones, word of mouth, local level announcement via mike and loud speakers
Spatial congregation of the disaster management organisations for quick meetings, planning and debriefing,
Toll free numbers 1070 and 1077 at the state and district operation centre
Continuous dialogue generate effective core information
World Views
Culture of disaster preparedness (building human and organisational capacities)
Learning culture
Improved Institutional arrangement for disaster management
Declaring 29 October as 'Disaster Preparedness Day'
Improving infrastructure for disaster management (bridges, schools, shelters, concrete houses, mobile)
Strategic leadership
Developing a 'goal' for 'zero casualty at any cost'
Agile bureaucracy
Reflection
Heightened awareness on gender and disaster preparedness
Compensation culture

conceived differently in some cases by the participants. An example is cited in this instance.

> The first death in Bhubaneswar occurred around 12.0 pm on 12th because the person came out of home to pluck flowers. A branch fell on her. That was the first death. How can we prevent such death? During cyclonic condition why she should come out to pluck flowers instead of remaining indoors.
>
> (Deputy Relief Commissioner, Special Relief Organisation, 22 July 2014, Rajiv Bhawan, Bhubaneswar).

The context of this death can be examined with a cultural lens highlighting the role of an individual's behaviour. Almost all Odiya households pray to their gods every morning. Household members, largely women and young girls, offer prayers, flowers and sweets to the gods. This is known as *Prasad* or an offering. Hence, the day that Phailin was to make a landfall was no exception. This particular individual came outdoors to pick flowers and it was during this act that the branch of a tree fell on her and caused her death. According to the Deputy Relief Commissioner, this was a self-inflicted death and not due to the response system. The disaster response system generated repeated weather warnings and alerts which were disregarded by this particular individual. This anecdote suggests that compensation is not enough; raising awareness, behaviour change information, education and communication materials, or trainings are equally vital to reduce deaths.

To conclude, a concerted effort was observed during the Cyclone Phailin to avoid deaths. These efforts were a process that began in the inter-disaster period between 1999 and 2013. During this period, the Government of Odisha developed a socio-technical disaster management system supported by the Government of India, UN, World Bank, INGOs, NGOs, media and the at-risk community. Investment in early warning systems, improved infrastructure, prepared at-risk community, decentralised policy and planning, sheer dedication of the responders and a strategic leadership exhibited by the Chief Minister of Odisha (amongst many)—led to a dramatic reduction of human deaths during the Cyclone Phailin. A compilation of these views are presented in the Table 4.2.

References

Ayan J, Basu R, Basu A (2015) Studying policy changes in disaster management in India: a tale of two cyclones. Disaster Med Public Health Preparedness 10(1):42–46. doi:10.1017/dmp.2015.116

Barik B (2014) Met Office goes digital. The Telegraph, 18 Feb

GoO (1996) The Orissa relief code. Revenue Department, Bhubaneswar

GoO (2001) Creation of Orissa disaster rapid action force. Home Department Resolution No 939/CD Dated 7 June 2001. http://odishapolice.gov.in/?q=node/208. Accessed 28 May 2015

GoO (2005) State disaster management policy: resolution 4th March 2005. Revenue Department, Bhubaneswar

Harriman L (2013) Cyclone Phyline in India: early warning and timely actions saves lives. UNEP Global Environmental Alert Services (GEAS). www.unep.org/geas/. Accessed 20 May 2015

Herrmann S (2009) Remote sensing. In: Glantz M (ed) Heads up!: early warning systems for climate, water and weather-related hazards. United Nations University Press, New York, pp 10–11
IANS (2013) Radio, mobiles saved lives during Phailin: survivors. Indo-Asian News Service, 20 Oct
Jenkins GM (1969) The systems approach. J Syst Eng 1:3–49
Lalonde C (2011) Managing crisis through organisational development: a conceptual framework. Disasters 35(2):443–464. doi:10.1111/j.1467-7717.2010.01223.x
Mishra BN (2013) Tryst with Phylin: the deadliest cyclone in 2013. The Response XIII (Oct 2013):5–7
Mudur GS, Mishra A, Mohanty S (2013) System shows its spine: battle 42 years in the making. The Telegraph, 16 Oct
Nayak AK (2013) A success story in preventing human causality in cyclone Phylin. The Response XIII (Oct 2013):8–12
OSDMA (2011) GoI-UNDP disaster risk reduction programme in Orissa: an overview. OSDMA, Bhubaneswar
OSDMA (2012) GoI-UNDP disaster risk reduction programme in Odisha: annual report. OSDMA, Bhubaneswar
OSDMA (2014) State disaster management plan. Government of Odisha, Bhubaneswar. www.osdma.org/Download/State_Disaster_Management. Accessed 6 July 2016
Prime R (1992) Hinduism and ecology: seeds of truth. Cassell Publishers Ltd, New York
Ray-Bennett NS (2009) Caste, class and gender in multiple disasters: the experiences of women-headed households in an Oriya village. VDM Verlag, Saarbrucken
Ray-Bennett NS (2010) Multiple disasters and micro-credit: can it reduce women's vulnerability?: the case of Tarasahi in Orissa, India. Disasters 34(1):240–260. doi:10.1111/j.1467-7717.2009.01127.x
Senge P (1990) The leader's new work: building learning organizations. Sloan Manag Rev 32 (1990):7–23
Srivastava SK (2009) Making a technological choice for disaster management and poverty alleviation. Disasters 33(1):58–81. doi:10.1111/j.1467-7717.2008.01062.x
UN (2015) Sendai framework for disaster risk reduction 2015–2030. UNISDR, Geneva. http://www.unisdr.org/we/coordinate/sendai-framework. Accessed 24 June 2016

Chapter 5
Systems Failure Revisited

Abstract This chapter summarises the previous chapters and outlines some of the limitations of systems failure. In doing so, this chapter proposes an emerging tool called 'systems alignment'. Perfect systems alignment is neither proposed nor is it possible in this real world because of people's subjective world views, different frames of reference, unique communication structures, cultures and the like. However, a weaker version of systems alignment, or some accommodation over the issue of reducing death can also lead to positive outcomes, as observed in the case of Cyclone Phailin by designing an overall goal of 'zero casualty at any cost' for the disaster management system. This chapter also outlines some organisational learnings that can take place from the findings of this research as well as some impact for policy and theory change in Odisha and beyond.

Keywords Avoidable deaths · Complex perspective · Coordination · Communication sub-systems · Decision making · Disaster management practices · Hazard · Human and organizational capacities · Justice · Policy · Political leadership · Systems alignment · Theory · Unavoidable deaths · Worldviews

It is argued that most of the deaths in disasters are avoidable. Avoidable deaths are preventable deaths due to advancements in disaster management science and weather forecasting systems; increased sophistication in human-built environments; as well as ongoing economic and policy development worldwide. When avoidable deaths continue to happen, this is event violence. Violence is commenced by the relevant actors and organisations in failing to protect or save precious lives. Deaths in disasters, in this vein, are a case for violation of justice. In the event of disasters, justice is denied to those women, men and children who would have otherwise lived a long life and an accomplished life. In order to promote justice in disasters, three arguments were made. First, human deaths must be identified as a matter of justice; as such they should receive a high priority from the disaster management system by developing a 'goal' to reduce death. Second, problems can be framed, as well as solved, within a disaster management system through support-led processes, such as effective INGO, NGO, community and government organisation

N.S. Ray-Bennett *Avoidable Deaths*, SpringerBriefs in Environmental Science,
https://doi.org/10.1007/978-3-319-66951-9_5

coordination and communication. Third, it can make room for demands of duty from the actors and organisations involved in protecting lives (Sen 1999, 2009) (Chap. 1).

By engaging with the existing risk and vulnerability perspectives, Chap. 2 aimed to explain why deaths in disasters occur. Risk or the traditional perspective enabled the advantage of understanding the dynamics of geohazards and their effect on humans. The vulnerability perspective on the other hand, helped in explaining why some groups of people are more vulnerable to disasters than others due to their class, gender, age, and race identities. An additional 'complex perspective' is also suggested to explain why deaths continue to occur. In this perspective, deaths occur due to the vulnerabilities that exist in the seams of disaster management system. This system is a conglomeration of different professional groupings and actors designed for specific tasks and goals. It is also a system that is highly reliant on technology. As such loose coordination and communication between actors can lead disaster management system to fail. To showcase how the disaster management system can fail to save lives, an analytical tool for systems failure was presented in Chap. 2. Systems failure has three inter-connected components: co-ordination, communication and world views. The analytical advantage of this tool was discussed in light of the two different case studies, Super-Cyclone of 1999 and Cyclone Phailin of 2013 in Chaps. 3 and 4. Table 5.1 exhibits the differences between the two disasters, as identified in the previous chapters while using the analytical tool of systems failure.

5.1 Systems Alignment

Systems failure provided a new perspective to analyse both the cyclones in Odisha. It also identified that systems' failures can be rectified by aligning the systems, which is coined here as 'systems alignment', in order to tackle wicked or complex problems, such as avoiding deaths in disasters. See Fig. 5.1 for the determinants of system alignment. Perfect systems alignment is neither proposed nor is it possible in this real world because of people's subjective world views, different frames of reference, unique communication structures, cultures and the like. However, a weaker version of systems alignment, or some accommodation over the issue of reducing deaths can also lead to positive outcomes, as observed in the case of Cyclone Phailin by designing an overall goal of 'zero casualty at any cost' for the disaster management system. Alignments for coordination and communication were achieved by investing in science and technology related to early warning systems, developing administrative and knowledge network and policy and planning—amongst many. World views were progressed through a culture of disaster preparedness, learning and a strong commitment to reduce deaths. A synopsis of these alignments of coordination, communication and world views in the context of Odisha/India is as follows for the purpose of learning.

Table 5.1 Difference between the Super-Cyclone and Cyclone Phailin (Produced by author)

	Coordination	Communication	World views
Super-Cyclone 1999	No coordination	Only via telephone	Complacent and conservative
	Core information available 48 h ago	Early warning system was under-developed	No goal
	Odisha Relief Code was activated afterwards	The communication system to disseminate information and warnings was under-developed	Lack of leadership
	Response system was reactionary	Limited forecast technology to generate core information	A fatalistic mind-set that hindered evacuation
	No authority to monitor relief and rescue		
	Lack of plan or planning		
Cyclone Phailin 2013	"excellent coordination" (IMD, Director)	Telephone, mobile, hot line, V-Sat, ham radio, fax, e-mail, radio, TV	
	Core information was available 4 days ago (>96 h)	Mike, loud speaker, word of mouth, force	Goal: "Zero Casualty at Any Cost"
	Knowledge and administrative networks developed by the Government of India helped analysing the core information	Early warning system was well developed	Strong, strategic and decentralised leadership
	Response system was proactive	The communication system to disseminate information was well developed	A change of mind-set with a heightened awareness on gender and disaster preparedness
	Numerous authorities/organisations monitored and assisted with rescue efforts	Warnings were disseminated very well, even by text and radio	Improved institutional arrangements for disaster management
	District Disaster Management was in place as well as the Disaster Management Act 2005	Improved infrastructure (road, cyclone shelters, public buildings)	
		New space-based technology the INSAT Satellite—3D	
		High power computer systems	

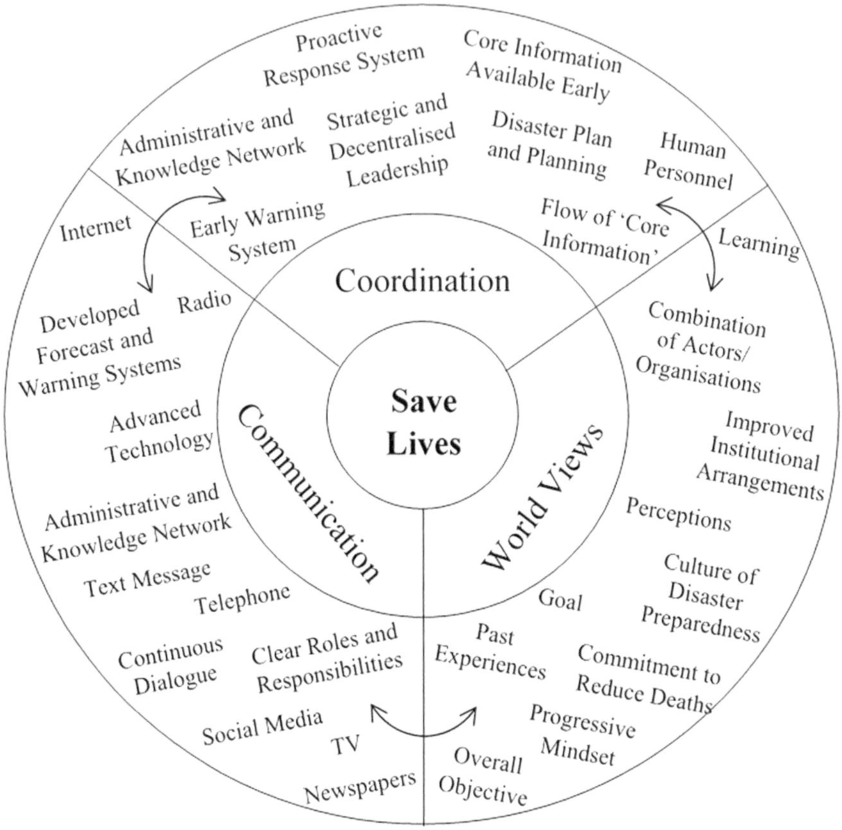

Fig. 5.1 Determinants of systems alignment to save lives (Produced by author)

5.1.1 Enhancing Coordination and Decision Making Processes

It was argued that when core information is unavailable, all actors are constrained in the coordination of the response. Accordingly, the availability and accessibility of this core information is crucial to rectifying the situation. In India and Odisha, the governments rectified this by investing in modern, up-to-date early warning systems. Investment in science and technology are important means of generating more core information and thereby, enhancing coordination (Alberts and Hayes 2003; UNISDR 2015b) as observed in the case of Odisha. These initiatives are consistent with the development occurring in the Asia Pacific region after the Indian Ocean Tsunami in 2004.[1]

[1]According to the international commitment through the Economic and Social Commission for Asia and the Pacific (ESCAP) Panel on Tropical Cyclones, the World Meteorological

The Government of India began the capacity building process by modernising the Indian Meteorology Department,[2] as well as India's disaster management system in the aftermath of the Super-Cyclone (Planning Commission 2013a). The decision to modernise the Indian Meteorology Department and the disaster management system was facilitated largely by two factors: economic growth and the recognition that India faces losses up to 2% of its GDP due to natural disasters. During the 10th (2003–2007) and 11th Five Year Plans (2007–2012), India's economy grew by 8% (Planning Commission[3] 2013a). In the 12th Five Year Plan (2012–2017), India is considered as the second fastest growing economy in the world. This success helped the central government to make strategic investments in science, technology, policy and planning to counteract the consequences of natural disasters. The strategies were translated into action by adopting two inter-connected components: *knowledge network* and *administrative network* (Srivastava 2009).

To develop the knowledge network, the Government of India established the Earth System Science Observation (ESSO) under the Ministry of Environmental Science in 2006. Earth System Science Observation observes earth systems to

---

(Footnote 1 continued)

Organization's Regional Specialized Meteorological Centre (RSMC-New Delhi) now issues information four to eight times a day to panel member countries when tropical cyclones form in the Bay of Bengal and the Arabian Sea (Pattie 2009). The panel member countries include: Thailand, Myanmar, Bangladesh, India, Pakistan, Sri Lanka, the Maldives and Oman (Pattie 2009). It is claimed that this core information made available by the World Meteorological Organization was vital to reducing human deaths during the cyclones Nargis in 2008 in Myanmar (Pattie 2009) and Mahasen in Bangladesh in 2013. The UN systems including the World Health Organization, United Nations Environment Programme, United Nations Development Programme, United Nations Food and Agricultural Organization, World Meteorological Organization, United Nations Educational, Scientific and Cultural Organization, United Nations International Strategy for Disaster Reduction, World Food Programme and The Famine Early Warning Systems Network – are now given a special charge by the aforementioned governments to foster awareness and develop the effective use of early warning systems in their institutional jurisdictions (Glantz 2009). In Sendai, the World Meteorological Organization announced more support to governments and others in developing multi-hazard early warning systems (UNISDR 2015b).

[2]Government of India modernised the Indian Meteorology Department by: '(i) commissioning 10 global positioning system stations; (ii) installing nine Doppler Weather Radars besides the existing five Doppler Weather Radars which have improved real time monitoring services; (iii) installing integrated Airport Meteorological Instruments at Mumbai, Hyderabad, Bangalore, Jaipur and Delhi airports; (iv) installing 550 Automatic Weather Stations apart from the existing 125, in addition to installation of 689 Automatic Rain Gauges; (v) commissioning of a set of four High Power Computers with a total installed capacity of 124 Teraflops for global data processing and Numerical Weather Prediction for weather forecasting services. A district-level agro-meteorological advisory service along with a five days in advance district-level weather forecast system, covering all the 555 districts, was launched for farmers in partnership with a number of Central Government ministries and organisations, state-level institutions, private agencies, non-governmental organizations, progressive farmers and the media. Over 3 million farmers have subscribed to receive this information through mobile phones' (Planning Commission 2013a: 254).

[3]The Planning Commission of India has been renamed as Nitiyog by the current Prime Minister, Mr. Narendro Modi.

understand the variability of these systems and to improve the weather forecasting system for hazards (Planning Commission 2013a). It also established India's Space Research Organisation (SRO) under which the Disaster Management Support (DMS) provides near real-time support in the form of imaging and communication satellites for efficient management of disasters in India (Planning Commission 2013a, b). The National Emergency Communication Network (NECN), a sub-communication system, has also been set up with the aim "to interconnect national, state and district Emergency Operation Centres (EOCs) as well as mobile and transportable Emergency Operation Centres which can be deployed at emergency and district situation" (Srivastava 2009: 70). These initiatives were pivotal in the reduction of human deaths during Phailin. Furthermore, the ongoing dialogue created as part of the knowledge network between the Indian Meteorology Department, Earth System Science Observation, India's Space Research Organisation, Central Water Commission, National Geography Research Institute and National Remote Sensing Organisation (Srivastava 2009) led to the generation of effective core information during Phailin, as well as its dissemination to the at-risk population (Director of Indian Meteorology Department, 21 and 22 July 2014, Bhubaneswar).

This national level development was matched with the modernisation of the Indian Meteorology Department in Odisha. The Indian Meteorology Department in Odisha was established in 1948 and was upgraded to a Cyclone-Warning Centre in 1973 after a severe cyclone in 1971 (Barik 2014). In 2002, the Government of India installed a Doppler radar in Paradeep (Indian Meteorology Department/Odisha 2014). Doppler radars monitor rain, wind and observe tropical cyclones (Kalsi 2003). The Indian Meteorology Department in Odisha also "monitors 21 observatories, 37 automatic weather stations, 177 automatic rain gauge stations, 149 rain gauge stations under different schemes, four pilot balloon observatories and 13 cyclone disaster mitigation committee stations across the state" (Barik 2014).

During the interview with the Director of Indian Meteorology Department, the Director said that the state government had agreed to install three additional Doppler radars in Gopalpur, Sambalpur and Baleswar in Odisha (21 July 2014; OSDMA 2011). This should enable them to provide weather updates every 15 min for thunderstorms, heavy rainfall and hail storms. Currently they work closely with the Indian Meteorology Departments in Kolkata, Vishakhapatnam and Patna to get weather updates every 4–5 h depending on the temporal conditions of the wind and cloud. As part of this knowledge network, the Director also mentioned that he interacts with the Indian Meteorology Department in New Delhi everyday via e-mails and telephone. This network between the different branches of Indian Meteorology Department was crucial during Phailin to generate effective core information, as well as coordinate this information across the relevant state disaster management departments in Odisha (Director of Indian Meteorology Department, 21 July 2014; Barik 2014). This knowledge network coupled with high power computing systems (Meteo France International Synergie System) enabled the Director and his Duty Officers to generate almost accurate early warnings during Phailin (Director of Indian Meteorology Department, 21 July 2014; Barik 2014).

According to the Director, their accuracy target is '85–90%' because forecasting can never be 100% accurate. Forecasting can only act as a 'reference'. However, this reference has to be infused with reflection, experience and wisdom of the practitioners. This can be explained in the words of the Director:

> Because of technology we are able to run the models in our office, we have servers, high-speed computers so it can run within 2 hours and generate forecasts of next 3–4 days. I can see this before 12 O'clock and give forecast at 12.30. So I got a reference, for which technology is crucial. Because human knowledge, human mind cannot predict more than 24 hours or 48 hours max. There is a limitation. But, we fully do not depend upon models. We do value additions. We exercise our experience, we add our observational data. But the reference helps. […] We get ideas. Then we look into cloud related weather patterns.
>
> **Author: When you say of value addition and experience, I understand that you have been in this job for a long time. Do you have any other practitioners working with you who are equally valuable in generating this knowledge collectively?** Oh yes. They are the Duty Officers […]. They will get trained for minimum 5 years. Each season has different characteristics. They face different types of weather each season. So while working for 5 years, they will learn what has been last year. What could happen this year or could be done or how was that and so on […].
>
> **Author: what is the academic background of this Duty Officers?** Earlier they used to have BSc in Physics or Chemistry or Maths. Or some came through promotion, but none through direct recruitment. But now the government has changed the recruitment policy. Duty Officers have to have postgraduates […] in maths, physics […] or meteorology or atmospheric science or geophysics or engineering. **Author: After 5 years, what happens to the Duty Officer?** Some will retire, some will become Sr. Duty Officers. […] We also have other groups such as Observational Groups. They are also scientific assistants […] but are much like entry cadre. They also had BScs like the Duty Officers […] but now this has been changed to postgraduates. […] They observe and whilst taking observations, they will understand what meteorology is? When they get promoted, they will be utilised in different shifts […].
>
> (Interview held on 22 July 2014, Indian Meteorology Department, Bhubaneswar).

The Indian Meteorology Department provided an ideal setting to observe the interplay of technology and human wisdom and knowledge. In the inter-disaster period, technology related to forecasting has advanced, and so has the demand for recruiting specialised human experts for these jobs. This new generation of experts will enhance the human-technology interaction, but wisdom and knowledge will still have to be gathered based on reflection-in/on-action and experience[4] (Schön 1983; Weick 1988). The reflective processes that the Duty Officers and the Observational Groups undergo as part of their training are necessary to build an *Auftragssystem*[5] Although *Auftragssystem* and 'power to the edge' are quite similar

[4]Reflection-in-action is based on a rapid interpretation of the situation (Ghaye and Ghaye 1998). It indicates that it commences in the middle of action, whereas 'reflection-on-action' occurs after the event to improve future action (Ghaye and Ghaye 1998; Schön 1983).

[5]This mission systems approach was developed by Adolph Hitler's military in Germany. The essence of this approach is that: "a subordinate commander, a subaltern […] should be trained to a level where he (or very rarely, she) could achieve the tactical goals of superior officers, with or without orders. […] Such a localised system of behavioural guidance makes heavy demands on the

concepts, the latter highlights that information communication tools are important to empower the responders working on the edge. Both of these concepts are useful in supporting the flow of core information in order to develop an effective disaster response system (Reason 1997) or a decentralised localised system so that they are capable of making critical decisions (Grint 2008). However, localised systems are dependent on the quality of the training received from first line of supervisors (Reason 1997).

At the Indian Meteorology Department, it is the Director who performs the supervisory role. He is an Officer from the Indian Administrative Service (IAS).[6] IAS is part of the Indian Civil Service which was established by the British government in colonial times. Currently the Civil Service forms the backbone of Indian government machinery. In addition to IAS, the Indian Civil Service recruits two other types of Officers: Central Services and State Services.[7] These Officers constitute all the major departments which run the state and central administration (UPSC 2015). The incumbents are part of an elite bureaucracy (Benbabaali 2008) and they play a crucial role in preserving national and state unity, integrity and uniform standards of administration by adopting a non-political, non-sectarian, secular outlook (GoI 2008). In the hierarchy of the Indian Civil Service, IAS officers are considered the 'cream' of India (GoI 2008). In this light, the current Director of Indian Meteorology Department is an outstanding individual. He also has a PhD in Meteorological Science. He took up his position in Odisha in 2007. Before that he was posted in Shillong, a north-eastern state of India. Although the author was unable to interview any Duty Officers, his presence, knowledge and experience made a huge contribution during Phailin as compared to the Super-Cyclone. The State Programme Officer of United Nations Development Programme is quoted in this regard:

> IMD (Indian Meteorology Department) played a crucial role. The information provided by IMD and Dr XXX was really precise and really concrete.

(Footnote 5 continued)

personal qualities of the superiors. A prerequisite is an extensive experience of the jobs carried out in the workplace of the conditions under which they are likely to be performed. […] Effective teams capable of operating autonomously when the circumstances demand it, need high quality leaders. This is in turn requires with the organisation invest heavily in the quality, motivation and experience – its first line supervisory" (Reason 1997: 218).

[6]The IAS exam is considered to be one of the toughest exams in India and the success rate is just 0.1% (GoI 2008; UPSC 2015).

[7]Author's other respondents including the three District Emergency Officers, Deputy General Manager of Odisha State Disaster Management Authority and the Deputy Relief Commissioner were recruited through the Odisha State Civil Service exam. Besides this, the District Emergency Officers of Jagatsinghpur and Ganjam had M.Phil. from two prestigious Universities of India (Jawaharlal Nehru University and the Institute of Population Studies in Mumbai) and the Deputy General Manager of Odisha State Disaster Management Authority had a Ph.D. All the respondents were also highly educated.

Developing Policies and Planning for Coordination: To promote an administrative network, the Planning Commission (2013b) outlined several strategies in the 10th Five Year Plan (2003–2007) which included:

i. policy guidelines on preparation of developmental plans;
ii. a multi-pronged strategy for risk management; and
iii. the recognition of a need for planned expenditure on disaster management and preventive measures in addition to the National Calamity Relief Fund.

These strategies were aligned with the UN's Hyogo Framework for Action, which the Government of India ratified in 2005 in Hyogo, Japan. The Government of India also passed the first ever 'Disaster Management Act 2005'- mentioned in the previous chapters. According to this Act, all the Indian states are required to develop a State Disaster Management Policy and Disaster Management Plans in order to decentralise the disaster management system[8] (OSDMA 2012). To facilitate the administrative network further, the Government of India also set up the National Institute of Disaster Management (NIDM) under the Ministry of Home Affairs in New Delhi. National Institute of Disaster Management builds the capacity of state administrators and responders through training and support. It is also geared towards providing support to the planners and practitioners through research (Srivastava 2009). To solidify this administrative network, the Government of India dedicated the 11th Five Year Plan (2007–2012) to disaster preparedness. The 12th Five Year Plan (2012–2017) is also designed to complement the previous initiatives by:

i. setting up early warning systems in all hazard prone areas of India;
ii. building communication networks;
iii. mainstreaming DRR into development activities; and
iv. building the capacity of the at-risk communities and of disaster management organisations.

For this, Indian Rupees 52,839 Crore (approx. GBP 5,562 Million) has been earmarked for the Ministry of Home Affairs (Planning Commission 2013a).

This wider policy change in India also coincided with the state building processes that began under the leadership of Mr. Naveen Patnaik in 1999. Under his direction, the Orissa Sate Disaster Mitigation Authority (later renamed as Odisha State Disaster Management Authority) was established, which helped immensely in co-coordination with local, national and international NGOs and multinational organisations, as well as towards the implementation of disaster preparedness and mitigation measures in the post Super-Cyclone era (GoO 2002; Ray-Bennett 2009; Samal 2003). Odisha State Disaster Management Authority and United Nations Development Programme Office together played a crucial role in building the capacities of the state, district, block and panchayati responders, through training and consultation (GoO 2001; OSDMA 2011). Capacity building training is critical

[8]Until 2005 only Odisha, Gujarat and Maharashtra had state disaster management policies in India.

in developing an effective localised *Auftragssystem* (Prizza 2007; Reason 1997). State Disaster Management Authority has also initiated NGO-GO coordination in the aftermath of the Super-Cyclone. The author had the privilege of attending several of these meetings during the floods of 2003 (Ray-Bennett 2009). Currently the NGO-GO coordination comprises of 22 INGOs, NGOs and UN agencies- together they are known as Inter Agency Group (OSDMA 2011). The Inter Agency Group played a crucial role in disseminating core information during Phailin (Deputy Relief Commissioner, State Programme Officer of United Nations Development Programme, Director of Indian Meteorology Department, 21–23 July 2014, Bhubaneswar). For instance, preparedness meetings were held among various Disaster Response and Volunteer Teams by the International Federation of Red Cross to assist with evacuation and relief (Harriman 2013). Also, State Disaster Management Authority and United Nations Development Programme Office together played a crucial role in building the capacity of the Inter Agency Group, as well as women's micro-credit groups through training and consultation (OSDMA 2011).

In light of these activities, Odisha State Disaster Management Authority is a vital thread that connects and coordinates with the primary, secondary and tertiary organisations and the at-risk communities in Odisha before, during and after disasters. Central to this coordinating process is the strategic positioning of Odisha State Disaster Management Authority. As discussed in Chap. 2, disaster management is a linear system, one in which organisations and actors are spatially separated (Perrow 1999). The Government of Odisha has managed to reduce this spatial separation by positioning Odisha State Disaster Management Authority with the other relevant government departments in one office complex. For instance, the State Disaster Management Authority's office is located a floor above the Special Relief Organisation and the Managing Directors of these organisations sit on the sixth floor of the same building. During the author's fieldwork, the roles of the Managing Director of the Odisha State Disaster Management Authority and the Special Relief Organisation were filled by the same person. This allowed the Director to align the disaster related policies and programmes of these two organisations. However, the Indian Meteorology Department was not located in the same office complex because it required space to host a number of technologies and equipment.

Nevertheless, the distance between Odisha State Disaster Management Authority and Indian Meteorology Department was less than 5 km. The close proximity of the relevant organisations means that the government actors are able to meet often, dine and discuss, talk over tea, and also arrange meetings at a very short notice. This was observed whilst dining with the members of Odisha State Disaster Management Authority, which included the Deputy General Manager. These congregations enhanced agility and communication between the departments and actors for the transmission of the core information prior to Phailin—something that was not possible in 1999. This strategic positioning of Odisha State Disaster Management Authority is then also conducive to promoting both knowledge and administrative networks with regard to disaster management.

5.1.2 Aligning Communication Sub-systems

Overcoming physical disruptions and levelling hierarchies that hinder communication is central to improving the communication sub-system (Chap. 2). Disaster management actors collectively are often spatially separated, as discussed above. They are also spatially separated from the at-risk population. Hence, maintaining the flow of this core information is vital. As mentioned earlier, information communication tools are increasingly used to communicate the core information. Yet, the breakdown of these tools is often inevitable, depending on the magnitude of the hazard (Comfort et al. 2004; Romo-Murphy et al. 2011). Accordingly, primary and secondary organisations must prioritise the securing of communication systems. A number of initiatives are already under-way with the help of space-based technology in this regard.

Space-based technology is considered as a major enabler for disaster management in Europe and the United States.[9] In India too, space-based technology, such as satellite phones and deployable VSAT terminals are now a major enabler for disaster management (Srivastava 2009). More recently, social media tools, such as Facebook, Twitter, Instagram, YouTube and blogs have demonstrated significant value during emergencies for the responders, at-risk population and for bystanders. These digital tools have been proven to give affected "communities a flexible information platform to share local knowledge, transparently document efforts, crowd-verification or eliminate false rumours, and engage in two-way communication with formal emergency response agencies" (Moore and Verity 2014: 3). For example, during the super typhoon Haiyan in the Philippines, Twitter enabled the first responders to engage with the public in real time and this interaction had proven to be very effective for emergency response planning.[10] To facilitate the increasing popularity of social media, the UN's Office for the Coordination of Humanitarian Affairs (OCHA) have standardised three hashtags for emergencies in 2014[11] (see Moore and Verity 2014; MacLean 2015). In Odisha, the role of social media was not emphasised by the responders. Mobile phones, hotlines, e-mails,

---

[9]Europe's Meteosat Generation Satellite (MSG) and the American National Oceanic and Atmospheric Administration (NOAA) are both responsible for weather forecasting (DLR 2013). As part of the International Charter, 'Space and Major Disasters' DLR's Centre for Satellite-Assisted Crisis Information (ZKI) in Germany played a crucial role in supplying up-to-date satellite images (with the help of TerraSAR-Z and RapidEye) to the Japanese government in the aftermath of the Japan Tsunami in 2011. The satellite images led to the development of maps in order to supply relief forces with useful information immediately (DLR 2013).

[10]"Within the first 48 hours after Super Typhoon Haiyan's landfall, nearly 230,000 tweets were published internationally containing a situationally relevant hashtag. From those tweets, over 600 written messages and 180 images were identified containing actionable information for emergency response planning" (Moore and Verity 2014: 3).

[11]Three hashtags for emergencies are: Early standardisation of the disaster name (e.g., #Fay), how to report non-emergency needs (e.g., #PublicRep) and requesting emergency assistance (e.g., #911US)" (Moore and Verity 2014: 2).

landlines, HAM radios, V-Sat, fax, radio, TV, word of mouth, community level announcement via the village task force—played an important role to maintain inter-organisational coordination, as well as coordination with at-risk community in disseminating the core information.

Disaster management in India (and elsewhere) is largely a government business and government offices and departments are hierarchical and bureaucratic in nature. Rigid hierarchical structures are detrimental to the flow of core information. These structures also promote a rigid mind-set. However, in this information age, these organisations are also under constant public gaze due to the burgeoning media industry and the pressure from NGOs and the UN (Director of the Indian Meteorology Department, 21 July 2014, Bhubaneswar). As a result, the standard operating procedures for a bureaucracy are in flux, and even more so during the dynamic phase of a disaster. In such a context, it was observed that the Government of Odisha leveraged the principles of 'command and control[12]' (C2), 'command, control and communication' (C3), and *Auftragssystem* or a localised disaster management system in order to maintain the flow of core information. Once the core information is generated by the Meteorology Department, Odisha State Disaster Management Authority, The Special Relief Organisation and the District Emergency Officers take ownership of this information and design their responses accordingly for the at-risk population. However, the ability of the actors to generate, disseminate and own the core information as well as develop the requisite response is embedded in the context of *Auftragssystem*. The 'Disaster Management Act 2005' has facilitated this *Auftragssystem* through the devolution of the disaster management system by creating a three-tier administrative structure (national, state and district) as well as Disaster Management Plans for the district, block and village levels.

United Nations Development Programme and Odisha State Disaster Management Authority together played key roles in developing these Plans in 30 districts. These Plans include specific instructions from Odisha State Disaster Management Authority, Special Relief Organisation and District Level Natural Calamity Committee. These Plans are living documents and are updated annually by the District Collectors in association with the Superintendent of Police, Assistant District Magistrates, Emergency Officers and the Line Department Officers. They are updated according to the requirements set by Odisha State Disaster Management Authority and United Nations Development Programme on issues relating to resources, human resources, technology and coordination (GoO 2012, 2014a).

[12]Command and control (C2) is different from the 'chain of command' approach (Alberts and Hayes 2003). The chain of command (C) is a traditional approach where the commander (traditional leader—see Senge 1990) is in charge of decision making. Command usually subsumed control in this approach. In C2, on the other hand, the command and control responsibilities are shared. There is no longer one commander in charge rather a collection of individuals (primary and secondary responders) who are assigned to accomplish a mission. Actions, inactions, active errors, cultures and mental models are of equal importance in the operation of C2 (Alberts and Hayes 2003: 14–15).

The District Disaster Management Plans have also led to a shift from a relief centric disaster management system to a pro-active disaster management system in the districts by prompting: (i) disaster risk assessment and vulnerability analysis; (ii) the identification of disaster prone areas; (iii) the identification of response structures; (iv) developing inventories of resources; and (v) developing standard operating procedures and incident command systems specific to each district (GoO 2013; OSDMA 2011).

The purpose of these District Disaster Management Plans is to help the district administrations (primary responders) and their associated line departments to focus quickly on the essentials of coordinating not just the core information but also decision making, preparedness and response (GoO 2013). These Plans have also led to the development of District Disaster Management Committees (DDMC) which are planning bodies with regard to preparedness and mitigation. District Disaster Management Committees help the District Collector in a disaster climate by: (i) reviewing the threat of disasters; (ii) assessing vulnerability; (iii) evaluating preparedness measures; and (iv) eliciting suggestions for the improvement of District Disaster Management Plans (GoO 2013: 19).

District Control Rooms/District Emergency Operation Centres (DEOC) are set up to align with the State Emergency Operation Centre (SEOP) and the Control Room at the Special Relief Commissioner's Office in the Special Relief Organisation in Bhubaneswar. The State Emergency Operation Centre is equipped with a state-of-art communication network (GoO 2015) in order to remain connected with the National Emergency Communication Network (NECN). Satellite phones and deployable VSAT terminals are major enablers for this (Srivastava 2009). Toll free numbers, 1070 and 1077 have also been installed in State and District Emergency Operation Centres (OSDMA 2011).

The operation of District Emergency Operation Centres is guided by command, control and communication (C3) approach (GoO 2013). As a hub of C3, the District Collector (commander): (i) reviews, monitors and analyses the potency of a hazard along with the District Disaster Management Committee (DDMC); (ii) co-ordinates the relevant line departments; (iii) deploys senior officers to vulnerable locations for rescue and evacuation and establishing a community kitchen; (iv) shares information continuously with the Control Rooms at the Revenue Department, Special Relief Organisation and Odisha State Disaster Management Authority in Bhubaneswar; and (v) implements appropriate actions and activities as part of the disaster management system (GoO 2013: 20). In a disaster climate, this hub becomes an important means of providing direction to the block and village level officials, NGOs and volunteers (GoO 2013). It is also noted that in a disaster climate C3 is an appropriate management style, whereas in 'normal' times it might not be. This command system ensured that the response to Cyclone Phailin was effective. It guided the primary responders' actions and facilitated their sense making processes in an uncertain situation. The outcome of this was an evacuation process that was unprecedented and a dramatic reduction in human loss.

Lastly, the early warning systems that generate core information must be seen as a vital component of disaster management through the lens of a systems approach

(Glantz 2009; Herrmann 2009). When there is a disconnect between the two, responses will be inadequate. This was observed during the drought in the Horn of Africa in 2011 (Aalst et al. 2013; Global Emergency Group, undated). Although the Famine Early Warning Systems Network (FEWS Net) forecast drought in the Horn of Africa about seven months prior to the disaster, the disaster management systems in the at-risk countries took no actions to prepare. Continuous coordination and communication between early warning systems and disaster management systems should enable a systematic response.

Frames of Reference: It is now known that the disaster management system is a conglomeration of different professional groupings and actors. It is also a system that is highly reliant on technology, as we have observed above. Also, it is observed that actors within this socio-technical system adopt different frames of reference. In this light, weak forms of organisation of the actors could potentially lead to systems failure (Chap. 2). Therefore, it is pivotal that the interconnections between organisations and actors are fully acknowledged. One way to do this is to synchronise early warning systems. Government, non-government and popular media use different forecasting terminology when warning the public and private businesses (GIZ 2012, 2015; Herrmann 2009). The wide range of warning schemes is confusing for the general public, and also makes it "difficult to disseminate the warning and initiate an effective evacuation and adequate response" (GIZ 2015: 1). To minimise this, it is important that the early warnings are 'people centred' by 'incorporating local practices' (UN 2005, 2015).

Despite this there are some success stories that have been documented, such as the Radio Djati initiative in Banda Aceh after the Tsunami in 2004 (Romo-Murphy et al. 2011). Radio Djati, a local initiative in Indonesia, caters to disabled and vulnerable groups. It has developed some essential equipment and procedures needed to operate as a rapid-response radio unit in future disasters. Radio Djati has also developed a consistent disaster information system regarding early warnings in collaboration with the state radio network of Indonesia, Radio Republic Indonesia and other stakeholders in the province of Aceh (Romo-Murphy et al. 2011).

In 2008, the International Federation of Red Cross pioneered a people centred early warning system in Mozambique called Early Warning Early Action (EWEA) (Braman et al. 2008). EWEA anticipates by making use of seasonal climate information at different time scales. The communication network is central to EWEA in order to develop linkages between early warnings and actions at the community level. This approach helped to save thousands of people's lives when the Barge Dam in Ghana spilled (see Braman et al. 2008).

In the lead up to the UN's Third World Conference on Disaster Risk Reduction in Japan in March 2015, GIZ (2015) proposed 'harmonisation of early warning alert levels'. This could prove beneficial much like the standardisation of the emergency hashtags for Twitter generated by the Office for the Coordination of Humanitarian Affairs (OCHA). GIZ suggested the harmonisation of the early warnings by adopting consistent colour codes (e.g. green, yellow, orange, red, for 'increasing dangers', followed by blue for an 'all clear status') and pictograms (GIZ 2015: 1). 'Alert levels in a standardised system' can be quickly recognised by the public much like the traffic light system. It is suggested that this harmonisation process has the potential to reduce confusion and align the communication sub-system.

In Odisha, the Special Relief Organisation issue a weather alert and the Indian Meteorology Department issue a weather warning. This division of labour helped the responders to facilitate their coordination and communication process in order to develop an effective response system.

5.1.3 World Views

In the previous chapters, the subjective world views of disaster risk reduction, gender, early warning and disaster management practitioners were examined. Harmonisation of the world views of these actors (science/scientists, social science/scientists, practice/practitioners, and lay science/lay people) is difficult to achieve due to the differing epistemological and ontological positions of these disciplines (Irwin 1995; Irwin and Wynne 1996; The Royal Society 1992). Nevertheless, it is suggested that *accommodation* of the world views could be a way forward. The process of accommodation will provide a space to host different world views or different points of view, as well as accommodate new ideas. This space is vital to resolving differences and seeking accommodation rather than just solutions (Checkland 1985). However, in some occasions accommodation might be difficult to achieve because human systems are highly complex and political systems (Vickers 1983). Some 'powerful' actors who passionately believe in a particular world view could go to extreme lengths and use all power resources to defend and advocate their views.[13] Thus, accommodation will not be easy; rather the dominant world view that of the powerful will prevail.

---

[13]In a visit to the Royal Society in London in February 2014, the Special Representative of the Secretary-General for disaster risk reduction, Ms. Margareta Wahlström announced 'that the second phase of the Hyogo Framework' (now the Sendai Framework) will be geared towards making Disaster Risk Reduction 'trans-boundary' by taking 'systems view'. The focus will also be on building organisational resilience, in addition to building the resilience of at-risk communities (Kemp 2014). The author was extremely excited by this announcement. On 24 June 2015, Margareta Wahlström revisited the Royal Society of London again as part of the Policy Lab Meeting on 'From Agreement to Action: What Next for the New Global Framework on Disasters?' The author attended this meeting. Margareta Wahlström provided an excellent synopsis of the Sendai Framework, one in which she recurrently emphasised the role of science in order to improve the disaster risk reduction practice.

After the meeting, the author spoke to Margareta Wahlström and reminded her of her earlier allusion to a systems view in February 2014. This is the summary of her response: 'There were not enough systems thinkers to negotiate the consultation phase of the Sendai Framework in Geneva. Rather the emphasis of DRR moved to science. They were strong. Every word of the Sendai Framework is owned by someone or somebody in this world. I think, systems thinking is now implicit in the Sendai Framework rather than explicit. However, this should not stop anyone pursuing research on this topic'.

Should we concede that an accommodation between actors is possible, then this has to begin by *acknowledging* the fact that human deaths in disasters are *avoidable deaths*. Second, intrinsically related to this, is the *recognition* by all actors that avoidable deaths are a case of 'event violence'. Third, to reduce such violence against women, men and children it is suggested that all actors and organisations working at the seams of disaster management systems should adopt an 'overall objective' or a goal (Jenkins 1969) which is to prevent deaths. Fourth, the translation of this objective has to be realised based on 'rational decision making model' (Pfeffer 1981). However, this process is not linear and will be fraught with power and politics. During the Cyclone Phailin, the Government of Odisha adopted the goal of 'zero human casualty' or the goal of unavoidable deaths (Chap. 4). If such a goal is ever achieved it is an ideal that illustrates the successful alignment of the human or socio-technical systems.

The Government of India, United Nations Development Programme Office and the Government of Odisha together played key roles in ushering a culture of disaster preparedness in Odisha. The Disaster Risk Management Programme, in particular helped: (i) by strengthening the state disaster management and district disaster management authorities; (ii) sharing knowledge and information to strengthen disaster management practises; (iii) mainstreaming disaster risk reduction into policies, plans and programmes; (iv) helping to prepare a disaster vulnerability atlas and Disaster Management Plans; (v) strengthening partnerships with various knowledge institutions for building the capacity of various stakeholders in DRR; (vi) building the capacity of vulnerable communities with regard to preparedness and response; and (vii) strengthening the governance arrangements for disaster response and preparedness through training (OSDMA 2012).

The culture of disaster preparedness also involved building the infrastructure required for disaster management, including good road conditions, bridges, concrete houses, multi-purpose cyclone shelters, schools and the like. Odisha State Disaster Management Authority in particular, has supervised the construction of 180 multipurpose cyclone shelters as well as community-based Shelter Management and Maintenance Committees (OSDMA 2011, 2012). It has also promoted disaster education in schools which are located within 15 km of the coast. About 96 schools have added disaster management to their curriculum (OSDMA 2011). The creation of the Odisha Rapid Action Force (ORAF) to support the district disaster response system, and plethora of community based disaster preparedness programmes by the NGOs and INGOs all contributed to build a culture of disaster preparedness.

Leadership is central to promoting a culture of disaster preparedness. Mr. Naveen Patnaik, the Chief Minister of Odisha provided much needed leadership in the aftermath of the Super-Cyclone. Since coming to power in 1999, he has declared 29 October as a 'Disaster Preparedness Day' for the state of Odisha in order to commemorate the Super-Cyclone. Since the year 2000, this is now an annual event celebrated throughout the state. Mr. Naveen Patnaik presides over this Day in Bhubaneswar and the event is attended by representatives of UN bodies, DFID, NGOs, students, volunteers and district and block level authorities. This event is often used as a venue to launch new initiatives (such as the 'Women in Disaster Preparedness and Mitigation' or 'volunteers in disaster management'), a platform to share successes and a medium to reflect collectively for future eventualities—amongst many (OSDMA 2002a, b).

5.2 Limitations

Having discussed some of the analytical advantages of systems failure through systems alignment, this section presents some of its limitations.

First limitation: This research focussed on disaster management organisations and did not engage with the at-risk communities in Odisha. This is partly because the ability to include subjects from social systems is rather limited in systems analysis (Mingers 1980). A systems thinker takes a 'privileged position' (Latour 2005) to design systems, systems problems and systems solutions (Dekker 2006; Dekker et al. 2011). This is because systems are ideas and the imagination of human minds. In this process of framing, many questions related to social systems prevail in this research: How do the actors of social systems report deaths? Or how don't they? What motivates death reporting? What hinders death reporting? Who reports death and how? Why deaths happen despite early warnings? In this light, systems failure and systems alignment will continue to maintain and reinforce 'the political status quo' of social systems (Jackson 1991; Mingers 1980). If one believes "that problems within society are not merely contingent but systematically created by the very structure of society then small scale problem-solving, 'piece-meal engineering', cannot help but maintain that which is the very problem—society itself" (Mingers 1980: 11). However, it was posited changing the 'very problem—society itself' is an impossible task. In this imperfect society, we would still require some solutions to reduce violence and demand for justice (Sen 2003).

Sen's idea of justice is succinct to argue for event specific violence or event violence in the context of this research. Some injustices are so gross, severe and intolerable, such as deaths in disasters that the actors and sympathisers from the human activity systems cannot wait for the entire society to change, rather take rapid actions to save lives. Event violence denies structural violence or a continuum of violence that exists in vulnerable groups of people's everyday lives (Cockburn 2004; Farmer 2004; Galtung 1969), which is consistent with Mingers argument mentioned earlier. Nevertheless, this prism of event violence also offers a vantage point from which a disaster response system can operate by developing goals and targets to save precious lives. Such an approach has the potential to give agency to the poor, vulnerable and the marginalised who may lack the ability to evacuate on their own or move to higher grounds prior to a disaster climate. The recognition of event violence must promote not only a meaningful disaster response system in many of the resource constrained developing countries but also promote accountability of the responders for the at-risk communities. It should also usher a culture of rights-based disaster management, one in which the at-risk communities can legitimately assert their rights to be rescued and evacuated to safe places prior to a disaster event. However, the recognition of event violence and the demand for structural change—two tenets of justice—are not contradictory. In the context of disaster management, they are complementary. Structural change can continue to take place in the mitigation phase, whereas the stopping of event violence can occur in a disaster climate. Both are mutually inclusive.

Second limitation: World views are important components of systems analysis. However, according to Mingers (1980), stressing the differences between world views leads to a surface level explanation rather than seeking explanation at the structures of society. This argument of Mingers stems from the idea of societal change or structural change as mentioned earlier. As a result, systems analysis lacks an explanation as to "why these particular Ws [world views] have developed and thereby how they might be changed. It lacks critical social theory" (Mingers 1980: 11). Furthermore, according to Mingers, outlining a possible world view is not enough because people can resist change. In this light, this approach lacks a theory in recognising the "difficulties of changing peoples' ways of thinking" (Mingers 1980: 11) or mental models. A snippet of this was visible in the cases of event violence in Odisha, such as the woman who came out to pick flowers on the day of Cyclone Phailin despite repeated weather warnings and alerts.

World views also generate several perspectives with regard to 'the problem'. Identifying a specific problem, as well as resolving this might be problematic in this approach (Checkland 1985). This is a paradox. In one hand, the solutions of wicked problems are inter-dependent, inter-disciplinary, inter-departmental, and multi-sectoral at the interface with technology. On the other hand, how can these be translated into actions. This is a mammoth task and could potentially be chaotic, confusing and also contribute to an ineffective disaster management system. At this point, many questions come to mind. Who will decide to seek wicked solutions? Who will coordinate such efforts? How will such efforts be coordinated? What motivates inter-departmental collaboration? How should such collaboration be facilitated? Who will facilitate and so on?

In the case of Odisha, it was the Chief Minister, who led such an effort. His efforts were facilitated by a plethora of policies, planning, programmes, funding, knowledge and administrative networks developed by the Government of India, UN, INGOs, NGOs, willingness of the at-risk community and the like. The case of Odisha illustrates the increasing role and involvement of political leadership before, during and after a disaster climate. When there is proactive political leadership, a disaster response system can be aligned with the goal of saving lives. Political leadership can promote a culture of disaster preparedness, too. In the case of Phailin, the Chief Minister set as a goal "zero casualty at any cost". Accordingly, all actors and responders organised themselves to achieve this target. The United Nations and other international funding organisations could do a great deal by encouraging political leadership to implement 'priorities for action' for effective disaster management (UN 2015; Ray-Bennett 2016).

Third Limitation: Theory testing and theory building are two crucial elements of a case study method (Eisenhart 1989; Yin 2009, 2012). According to Flyvbjerg

(2006), this process should happen in a real-world context in order to develop context-dependent knowledge or theory, as compared to context-independent knowledge. The latter approach is popular amongst neo-positivists with the aim to generalise research theories and outcomes. Context-dependent knowledge is relevant to this research because it identified the real life issues in disaster management systems with the aim of generating learning and, perhaps, a more relevant theory that can directly relate to the practices of disaster management authorities. One downside is that context-dependent knowledge can generate a very complex theory (Eisenhart 1989). This complexity comes from the voluminous amount of data that a case study inquirer encounters in the real world. Perhaps systems failure and its emergent component, systems alignment developed in this research, are no exception. At the outset readers might feel that the tenets of systems failure (co-ordination, communication and world views) are lengthy and complex. Said that, it is also important to note that complexity is at the heart of human deaths in disasters and so excessive simplification is likely to be counter-productive.

Fourth limitation: the focus has primarily been on developing an effective response system for cyclone/flooding. The availability of core information is vital to develop a response system for these hazards. In this context, systems failure presented is hazard specific. For some hazards such as an earthquake or tsunami, the generation and dissemination of core information might be problematic because the early warning systems are still evolving. Nevertheless, the components of systems failure have transferable value for disaster management in general.

5.3 Potential Impact

Despite its limitations, systems failure can lead to several organisational specific learnings, of which three are outlined here. First, organisations can improve disaster management practices and philosophy; second, they can develop human and organisational capacities in order to deal with contemporary disaster risks; third, it can inform policy making and theory development. They are all inter-connected in reality. Practitioners and researchers are also encouraged to improvise on the components of systems failure in order to suit the context of their own organisations and the hazards that they are exposed to.

Improve Disaster Management Practices: Managing disaster risks from the perspective of systems failure can benefit disaster management significantly (White 1995), as observed in the case of Odisha. A systems thinking promotes interdependencies and collaborations between actors and organisations in order to deal with risks of modern days. But interdependencies and collaborations cannot be achieved through science and technology alone (UN 2005, 2015). They have to be complemented by management knowledge, strategies, skills and reflection - amongst many. It is suggested that the components of systems failure can contribute significantly towards this.

Coordination and communication, in particular, are identifiable. They can help organisations to manage labour, knowledge and gaps in disaster management activities. They can also promote sense making amongst the organisations by questioning:

- What is 'it' that needs coordinating?
- How can coordination of 'it' be supported, advanced and facilitated by administrative structures, processes and technology?
- What communication techniques are required to enhance coordination between human to human, human to technology and technology to human interfaces?
- What is the goal of this coordination and communication?
- Who decides this goal?
- How can this goal affect the at-risk communities?

The best outcome of a goal will be achieved if the decision making process is based on the rational model, as compared to bureaucratic and political models.[14] Successful outcomes will also be facilitated if the alignment processes of actors and organisations have begun in the mitigation phase rather than in a disaster climate. Alignment of actors and organisations is a pre-requisite in a disaster climate. The successful achievement of a goal will also be influenced if the leadership of a response system is context-specific, organisational-specific and hazard specific. However, an overall goal to reduce deaths or to achieve zero casualties will have a far-reaching effect if this goal is set by the head of a state or by the head of a nation as compared to a bureaucratic head from a disaster management organisation in the event of a large-scale disaster.

Develop Human and Organisational Capacities: Building human and organisational capacity is pivotal to maintain the momentum of the responders. It is also important for continual maintenance of the disaster management institutions designed to serve a specific purpose for the human systems. However, the assumption that actors and organisations are well equipped to deal with the contemporary disasters risks, particularly in developing countries, is widespread. This assumption was promoted by the Hyogo Framework. The question remains: how can we best build the capacity of those organisations that apply the principles of disaster risk reduction? The Hyogo Framework placed much emphasis on the capacity building of the at-risk community compared to the organisations and professionals who are involved in the day-to-day messy businesses of averting disaster risks (Ray-Bennett et al. 2014a, b). As a result, the building of organisational capacity in national and local authorities, NGOs and environmental agencies has received far less attention by the disaster risk reduction community (DFID 2012, 2013). However, capacity building of the responding organisations is now

[14]The bureaucratic model, "rely more heavily on rules, precedent, and standard operating procedures. Less time and resources will be spent on decision making, and fewer alternatives will be considered before actions are taken" (Pfeffer 1981: 24). The political decision making model, presumes "that parochial interests and preferences control choice" (for details see Pfeffer 1981: 22).

one of the 'priority for actions'[15] for the Sendai Framework, which says: "empower local authorities as appropriate, through regulatory and financial means to work and coordinate with civil society […] in disaster risk management at local level" (UN 2015: 13). It is suggested that the model of socio-technical disaster management, as well as the tool of disaster climate can help building the capacity of not just local authorities but also of the associated organisations, including the secondary and tertiary.

Disaster climate is conceived in the context of systems failure. Disaster climate encapsulates uncertainties, dynamism, imagination, critical decision making processes, readiness, incident management and communication. It is suggested that the preparation for a disaster climate can lead to scenario planning[16] amongst the responders and in doing so, build the capacity of responders to detect blind spots, human errors and new possibilities (Senge 1990; Taleb 2007). Preparation for a disaster climate can also allow responders to expand their imagination and heighten awareness of a crisis (Perrow 1999; Senge 1990; Weick 1988). Heightened awareness should enable responders to see the 'bigger picture' in order to be prepared for a developing crisis. Training, consultation, debriefs and knowledge exchange can all contribute to the capacity building of responders and organisations. Other methods that can also heighten the awareness of the responders are Causal Loop Diagrams[17] (CLD), Outcome Mapping[18] (OM), and Assumption Based Planning[19] (ABP).

---

[15]'Strengthening disaster risk governance to manage disaster risk' (Priority for Action 2).

[16]Scenario planning (Masys 2012) is essentially a story, describing potential future conditions and their emergence to facilitate sense making and to inform decision making. The thought process involved in scenario planning supports 'thinking the unthinkable' exploring uncertainty and challenging mental models and assumptions in order to recognise alternate futures in a space of possibilities (VCLL 2013). There are numerous approaches to scenario planning in the literature such as Schwartz's 8-Step Scenario Building Model (see Schwartz 1996); Schoemaker's 10-Step Scenario Building Model (see Schoemaker 1993); Avin's 12-Step Scenario Building Model (see Avin 2007), and JISC's (2007) 6-Step process for the development of scenarios. By revealing the uncertainty one opens up the notion that more than one future is potentially open (VCLL 2013).

[17]Casual Loop Diagrams are important tool that can provide a language for articulating our understanding of dynamic, interconnected situations. Through the 'visual grammar' of Casual Loop Diagram, it prepares the participants to challenge the linear cause and effect relationships and decipher various interconnected feedback loops (VCLL 2013). Sterman (2000) and Senge (1990) and Senge et al. (1994) provide an excellent resource for learning about Casual Loop Diagrams (VCLL 2013).

[18]Outcome Mapping was developed by the International Development Research Centre (IDRC) in Canada in support of development efforts around the world (IDRC 2001). The application of this methodology to the disaster management domain is particularly relevant to the notion of building/enabling resilience. Two of the key features or principles of Outcome Mapping that resonate with systems thinking are its recognition of the importance of embracing different world views and perspectives as well as the acceptance of non-linear (complex) causality. The 12 steps associated with the Outcome Mapping methodology are available at IDRC (see 2001).

[19]Assumption Based Planning (ABP) is an essential element in disaster management domain according to Dewar et al. (1993). It is a powerful tool to show how plans often fail because inadequate attention was paid to the underlying assumptions. This tool aims to capture the

The current disaster management model is conceived as socio-technical. In this model, human and technology work at interface. A snippet of this model is visible in the seventh Global Target of the Sendai Framework: "substantially increase the availability of and access to multi-hazard early warning systems and disaster risk information and assessments to the people by 2030" (UN 2015: 7). Planning disaster management from the perspective of socio-technical systems will allow governments and international organisations to locate gaps and strengths in technologies (related to early warning systems and information and communication tools), as well as in human knowledge and expertise. This will usher a new type of disaster management practice which was observed in Odisha. This approach will also lead to the development of new research, new knowledge and novel ways of understanding the human technology interface to improve disaster response systems. Learning based on this new model must also extend to gender and disaster studies in order to develop gender sensitive theories, which mirror practice and the hyper connected global world in which vulnerable subjects are located.

Policy Impact: This research can also inform policy making. There is an urgent necessity to maintain gender disaggregated death records locally, nationally and globally for future research and development. There is, however, a move towards disaggregation due to the recommendation of the Hyogo Framework in 2005. Mortality data in disasters is something that neither the international organisations, such as Centre on the Epidemiology of Disaster (CRED), nor the national governments maintain (Eklund and Tellier 2012; Sanchez et al. 2009). In Odisha the problem is systemic.

It was noticed that disaggregated mortality data by caste is filtered out at the block office level, whereas class and gender data is filtered at the district office level.[20] Due to this practise, the Odisha State Disaster Management Authority and the Special Relief Organisation in Bhubaneswar, receive caste, class and gender neutral mortality data. Caste data is lost at the block office because the pro forma designed for the local government doctor to issue a death certificate due to disasters (primary death) does not require caste to be specified. Class and gender data is lost at the district office when it is sent by the block office for further verification. Death verifications are a serious matter because according to the Orissa Relief Code

(Footnote 19 continued)

assumptions underlying plans that if invalid, could derail the plans and operations stemming from it. It is a systemic tool because it explores the interrelationships, perspectives and boundaries of an established plan to ensure the viability of that plan. Through this process, it recognises how 'load bearing vulnerable assumptions' require explicit analysis (see Dewar et al. 1993).

[20]"Indian states comprise a three-tier administrative structure. Several *gram sansad* (villages) or wards (hamlets) constitute a *gram panchayat* (GP), several GPs constitute a *panchayat samiti* (PS) or block, and several panchayat samiti constitute a *zilla parishad* or a district" (Ray-Bennett 2009a: 12).

(GoO 1996), regardless of the deceased's caste, class or gender their next of kin is entitled to an ex gratia payment or a compensation. This compensation is hazard specific[21] and is approved by the District Collector (GoO 1996, 2014a, b). Consequently, the district and block offices are not required to collect disaggregated data.

The death reporting and recording system merits review in Odisha in light of the Sendai Framework. Recently, the Sendai's Policy Brief (Fakhruddin et al. 2017) developed by the International Council for Science and Integrated Research on Disaster Risk (IRDR), suggested "[e]stablishing basic data infrastructure for disaster loss data in developing countries and supporting regional and global cooperation for disaster loss reporting". In light of this, some concerted efforts are required to develop such infrastructure at regional, national, state and district levels in India and the SAARC region. It is also important to make the disaster loss data management consistent with the Guidelines identified by Integrated Research on Disaster Risk (2015) on 'measuring losses from disasters'. For this, the Government of Odisha should aim to document not only primary deaths, but also secondary deaths and those who are missing (Dilley and Grasso 2016; IRDR 2015).

Although it was beyond the scope of this research to capture people's perspectives on Odisha's death reporting system, it is likely that deaths could go un-reported—not by the beneficiaries but rather due to the verification process carried out by the government doctor at local government hospitals. This verification process merits further investigation. It is also important to address the death reporting practices in India in order to standardize this system, (Dilley and Grasso 2016; Fakhruddin et al. 2017) and introduce good practices in order to improve the accuracy of disaster death data. To be consistent with the Sendai Framework it is also recommended that the Government of Odisha record disaggregated data. This is not currently expected according to the Relief Code. In order to promote gender disaggregated mortality data, the UN's International Strategy for Disaster Risk Reduction and Integrated Research on Disaster Reduction can take a lead role, in association with the national governments and the United Nations Development Programme Office. In India, the National Institute of Disaster Management in New Delhi and the Odisha State Disaster Management Authority in Odisha, along with United Nations Development Programme Office in Bhubaneswar can undertake this venture. Longitudinal death records at local, national and global levels will be extremely important to indicate past and future death patterns in disasters. They will also promote necessary policy, planning and response by guiding how and where the finite resources of disaster risk reduction could be better placed to reduce human deaths or change human behaviors.

Theory Impact: Our knowledge on deaths or avoidable deaths in the context of disaster risk reduction is limited. Deaths in disasters are a developmental

---

[21]Indian Rupees 100,000.00 (approximately Great British Pound 1,025.00) for a death due to cyclone, Indian Rupees 150,000.00 (approximately GBP 1,500.00) due to lightening, Indian Rupees 10,000 (approximately Great British Pound 100.00) due to heat wave and Indian Rupees 100,000 due to a snakebite during floods (GoO 2014a).

issue. They result in both a humanitarian loss and a loss of human capital (Lass et al. 2011). Rural developmental studies have clearly noted the impact that a woman's death can have on livelihoods as well as on social, natural and political capital (Agarwal 1990; DFID 1999; Ray-Bennett 2009). Without reducing women and men's deaths, the WHO (2002) and UNDP (2007) have firmly asserted that it is unlikely that a country can ever achieve sustainable human development. This is increasingly evident in those countries which are repeatedly affected by disasters and consistently lagging in the achievement of the sustainable goals (UNDP 2007). Avoidable deaths, therefore, aims to provide an essential background material to support methods and theories that can better discern vulnerability and steps towards establishing solutions to reduce deaths.

Indicators for avoidable deaths were suggested in Chap. 1. Indicators both at global and national levels are important milestones or benchmarks. They act as targets to be achieved by the international, national and state disaster management authorities (UN 2015; Wahlström 2015). Some ideas as how to conceive these indicators through the instance of avoidable deaths and unavoidable deaths are suggested here. Anything less than ten deaths is considered unavoidable deaths. Avoidable deaths, on the other hand, are deaths beyond the number ten and are potentially avoidable from the impact of environmental disasters in the present time, given available knowledge on the nature of these hazards and due to advancements in information technology, human interaction and effective policy interventions. These indicators have both quantitative and qualitative elements. Depending on the economy of a country (developed, developing, least developed, middle income), indicators will vary, much like the Sustainable Development Goals.

Recently, the Sendai Framework has opened up an opportunity to initiate a global debate in this regard by setting a global target of reducing 'global disaster mortality by 2030'. UNISDR is in a unique place to lead this global debate to discuss the issues related to resources, processes and technology that will be required to build the capacity of the UN Member States to achieve this target. More research is therefore required to understand the contextual processes to reduce global disaster mortality. There is also a necessity to develop appropriate methodology for avoidable death indicators that can stand up to capturing the nuances of national and global contexts. In this regard, the disaster risk reduction community can benefit significantly from the methodological practices prevalent in the health and poverty reduction sectors. It is posited that the indicators for avoidable deaths in disasters will lead to the convergence of the Sendai Framework with the Sustainable Development Goals for a sustainable human development. This convergence is pivotal since disaster risk reduction is a development issue and likewise development is disaster risk reduction (UN 2015).

Last but not the least, conceiving 'deaths in disasters' through the lens of complex perspective should promote an accountable disaster management which is still in its nascent stage in Odisha. It should also help to identify gaps (such as no measures for lightning), incapabilities and inactions that often exist at the seams of human-built disaster management system. This is because the complex perspective allows us to redirect our attention to the actors' and organisations' actions and inactions to prevent deaths, rather than focussing solely on the severity of a risk and the vulnerabilities that exist amongst the at-risk community. When we are able to better comprehend the vulnerabilities of the disaster management system, it is more likely that we will be able to come up with concrete solutions and develop the skills, knowledge and expertise that will be required for the actors and organizations to reduce deaths of men, women and children.

References

Aalst MV, Kellett J, Pichon F, Mitchell T (2013) Incentives in disaster risk management and humanitarian response: background note for world development report. Overseas Development Institute, London

Agarwal B (1990) Social security and the family: coping with seasonality and calamity in rural India. J Peasant Stud 17(3):341–412

Alberts DS, Hayes RE (2003) Power to the edge: command, control in the information age. CCRP Publication Series. www.dodccrp.org. Accessed 6 Feb 2014

Avin U (2007) Using scenarios to make urban plans. In: Hopkins LD, Zapata M (eds) Engaging our futures: tools for effective planning practices. Lincoln Institute of Land Policy, Cambridge

Barik B (2014) Met Office goes digital. The Telegraph, 18 Feb

Benbabaali D (2008) Questioning the role of the Indian Administrative Services in national integration. South Asia Multidiscip Acad J. http://samaj.revues.org/633. Accessed 12 May 2015

Braman LM, Van Aalst MK, Mason SJ, Suarez P, Ait-Chellouche Y, Tall A (2008) Climate forecasts in disaster management: Red Cross flood operations in West Africa. Disasters 37 (1):144–164. doi:10.1111/j.1467-7717.2012.01297.x

Checkland P (1985) From optimizing to learning: a development of systems thinking in the 1990s. J Oper Res Soc 36(9):757–767. doi:10.1057/jors.1985.141

Cockburn C (2004) The continuum of violence: a gender perspective on war and peace. In: Giles W, Hyndman J (eds) Sites of violence: gender and conflict zones. University of California Press, California

Comfort LK, Ko K, Zagoreck A (2004) Coordination in rapidly evolving disaster response systems: the role of information. Am Behav Sci 48(3):295–313. doi:10.1177/0002764204268987

Dekker S (2006) The field guide to understanding human error. Ashgate, Surrey

Dekker S, Cilliers P, Hofmeyr JH (2011) The complexity of failure: implications of complexity theory for safety investigations. Saf Sci 49(6):939–945. doi:10.1016/j.ssci.2011.01.008

Dewar JA, Builder CH, Hix WM, Levin MH (1993) Assumption-based planning: a planning tool for very uncertain times. RAND Corporation. http://www.rand.org/content/dam/rand/pubs/monograph_reports/2005/MR114.pdf. Accessed 5 Jan 2016

Dilley M and Grasso V.F. (2016) Disaster reduction, loss and damage data, and the post-2015 international policy agenda. Environ Sci Policy 61:74–76. doi:10.1016/j.envsci.2016.04.002

DFID (1999) 'Framework', sustainable livelihoods guidance sheets. http://www.livelihoods.org/info/info_guidancesheets.html. Accessed 22 October 2005

DFID (2012) Minimum standards for embedding disaster resilience in DFID country offices. DFID, London

DFID (2013) Helping developing countries deal with humanitarian emergencies. DFID, London

DLR (2013) Earth observation: discovering, surveying, and understanding our planet. Deutsches Zentrum fur Luft-und Raumfahrt e.V (DLR), Bonn

Eisenhart K (1989) Building theories from case study research. Acad Manag Rev 14(4):532–550. doi:10.5465/AMR.1989.4308385
Eklund L, Tellier S (2012) Gender and international crises response: do we have the data, and does it matter?. Disasters 36(4):589–608. doi:10.1111/j.1467-7717.2012.01276.x
Fakhruddin B, Murray V, Maini R (2017) Disaster loss data in monitoring the implementation of the Sendai Framework [Policy brief]. https://www.icsu.org/cms/2017/05/DRR-policy-brief-2-data.pdf. Accessed 25 June 2017
Farmer P (2004) Pathologies of power: health, human rights, and the new war on the poor. University of California Press, California
Flyvbjerg B (2006) Five misunderstandings about case study research. Qual Inquiry 12(2):219–245. doi:10.1177/1077800405284363
Galtung J (1969) Violence, peace, and peace research, J Peace Res 6(3):167–191. doi:10.1177/002234336900600301
Ghaye A, Ghaye K (1998) Teaching and learning through critical reflective practice. David Fulton Publishers, London
GIZ (2012) LFEWS: local flood and early warning system. GIZ Publication, Manila
GIZ (2015) Harmonisation of early warning alert levels. GIZ Publication, Manila
Glantz M (2009) Introduction. In: Glantz M (ed) Heads up!: early warning systems for climate, water and weather-related hazards. United Nations University Press, New York, pp 1–28
Global Emergency Group (undated) IASC real-time evaluation of the humanitarian response to the Horn of Africa drought crisis, Kenya. Interagency Standing Committee, Virginia
GoI (2008) History of civil services in India and reforms. ARC 10th report. www.arg.gov.in. Accessed 12 May 2015
GoO (1996) The Orissa relief code. Revenue Department, Bhubaneswar
GoO (2001) Creation of Orissa disaster rapid action force. Home Department Resolution No 939/CD Dated 7th June 2001. http://odishapolice.gov.in/?q=node/208. Accessed 28 May 2015
GoO (2002) Orissa state development report. Nabo Krishna Centre for Development Studies, Bhubaneswar
GoO (2012) District disaster management plan 2012–13: Cuttack. Emergency Operation Center (EOC) Collectorate Office, Cuttack
GoO (2013) District disaster management plan 2013–14: Jagatsinghpur District. District Emergency Operation Center, Jagatsinghpur
GoO (2014a) District disaster management plan 2014–15: Puri District. Emergency Operation Centre, Puri
GoO (2014b) Ganjam disaster district management plan 2014–15: Chatrapur District Emergency Operation Center, Chatrapur
GoO (2015) About department. Revenue and Disaster Management Department. http://www.odisha.gov.in/revenue/web/Aboutus.asp?GL=1&PL=1. Accessed 2 June 2015
Grint K (2008) Wicked problems and clumsy solutions: the role of leadership. Clin Lead I(II):11–15
Harriman L (2013) Cyclone Phyline in India: early warning and timely actions saves lives. UNEP Global Environmental Alert Services (GEAS). www.unep.org/geas. Accessed 20 May 2015
Herrmann S (2009) Remote sensing. In: Glantz M (ed) Heads up!: early warning systems for climate, water and weather-related hazards. United Nations University Press, New York, pp 10–11
IDRC (2001) Outcome mapping: building learning and reflection into development programs. IDRC. http://www.idrc.ca/EN/Resources/Publications/Pages/IDRCBookDetails.aspx?PublicationID=121. Accessed 5 Jan 2016
Integrated Research on Disaster Risk (2015) Guidelines on measuring losses from disasters: human and economic impact indicators (IRDR DATA Publication No. 2). Integrated Research on Disaster Risk, Beijing
Irwin A (1995) Citizen science: a study of people, expertise and sustainable development. Routledge, London

Irwin A, Wynne B (1996) Introduction. In: Irwin A, Wynne B (eds) Misunderstanding science?: the public reconstruction of science and technology. Cambridge University Press, Cambridge, pp 1–18

Jackson MC (1991) The origins and nature of critical systems thinking. Syst Pract 4(2):131–149. doi:10.1007/BF01068246

Jenkins GM (1969) The systems approach. J Syst Eng 1:3–49

JISC (2007) Scenario planning. JISC. www.jiscinfonet.ac.uk/tools/scenario-planning.pdf. Accessed 5 Jan 2016

Kalsi SR (2003) Monitoring and forecasting of tropical cyclones and their associated effects. In: Gupta HK (ed) Disaster management. Universities Press, Mumbai

Kemp R (2014) Building within the scaffolding: The architecture of the new Hyogo Framework for Action. The Royal Society. http://blogs.royalsociety.org/in-verba/2014/03/24/new-hyogo-framework/?utm_source=social_media=hootsuite&utm_campaign=standard. Accessed 26 Feb 2014

Lass W, Haas A, Hinkel J, Jaeger C (2011) Avoiding the avoidable: towards a European heat waves risk governance. Int J Disaster Risk Sci 2(1):1–14. doi:10.1007/s13753-011-0001-z

Latour B (2005) Reassembling the social: an introduction to actor-network theory. Oxford University Press, Oxford

MacLean J (2015) Can standardised hashtags be effective in emergency responses?. Communications technology and security risk management blog. http://commstech-hub.eisf.eu/news—blog. Accessed 1 Apr 2015

Masys AJ (2012) The emergent nature of risk as a product of "heterogeneous engineering": a relational analysis of oil and gas industry safety culture. In: Bennett S (ed) Innovative thinking in risk, crisis, and disaster management. Gower, Surrey, pp 35–46

Mingers JC (1980) Towards an appropriate social theory for applied systems thinking: critical theory and soft systems methodology. J Appl Syst Anal 7:41–49

Moore R and Verity A (2014) Hashtag standards for emergencies: OCHA policy and studies brief. United Nations Office for the Coordination of Humanitarian Affairs (OCHA), Geneva

OSDMA (2002a) Managing disasters in Orissa: background, challenges and perspectives. OSDMA Publication, Bhubaneswar

OSDMA (2002b) Disaster preparedness day. The response. Q Newslett 11(3):1–12

OSDMA (2011) GoI-UNDP disaster risk reduction programme in Orissa: an overview. OSDMA, Bhubaneswar

OSDMA (2012) GoI-UNDP disaster risk reduction programme in Odisha: annual report. OSDMA, Bhubaneswar

Pattie D (2009) Cyclone Nargis. In: Glantz M (ed) Heads up!: early warning systems for climate, water and weather-related hazards. United Nations University Press, New York, pp 38–41

Perrow C (1999) Normal accidents: living with high-risk technologies, 2nd edn. Princeton University Press, Princeton

Pfeffer J (1981) Power in organization. Pitman Publishing Inc, London

Commission Planning (2013a) Twelfth five year plan 2012–2017: faster, more inclusive and sustainable growth: vol 1. Sage, New Delhi

Commission Planning (2013b) Twelfth five year plan 2012–2017: economic sectors, vol 2. Sage, New Delhi

Prizza R (2007) The role of coordination in disaster management. In: Pinkowski J (ed) Disaster management handbook. CRC Press, New York, pp 75–96

Ray-Bennett NS (2009) Caste, class and gender in multiple disasters: the experiences of women-headed households in an Oriya village. VDM Verlag, Saarbrucken

Ray-Bennett NS, Masys AJ, Shiroshita H, Jackson P (2014a) Hyper-risks in a hyper connected world: a call for critical "reflective response" to develop organisational resilience. Input paper prepared for the Global Assessment Report on Disaster Risk Reduction 2015. UNISDR, Geneva

Ray-Bennett NS, Masys AJ, Shiroshita H, Jackson P (2014b) 'Reactive to pro-active to reflective disaster response: introducing critical reflective practices in disaster risk reduction'. In: Collins AE, Samantha J, Manyena B, Jayawickrama J (eds) Natural Hazards, Risks, and Disasters in Society: A Cross-Disciplinary Overview, Amsterdam: Elsevier, pp 99–114

Ray-Bennett NS (2016) Learning from deaths in disasters: the case of Odisha, India. MEI Middle East-Asia Project (MAP) on humanitarian assistance and disaster relief: rising to the challenge?. MAP Bulletin. http://www.mei.edu/content/map/learning-deaths-disasters-case-odishaindia

Reason J (1997) Managing the risks of organisational accidents. Ashgate, Aldershot

Romo-Murphy E, James R, Adams M (2011) Facilitating disaster preparedness through local radio broadcasting. Disasters 35(4):801–815. doi:10.1111/j.1467-7717.2011.01234.x

Sanchez C, Lee TS, Young S, Batts D, Benjamin J, Malilay J (2009) Risk factors for mortality during the 2002 landslide in Chuuk, Federated States of Micronesia. Disasters 33(4):705–720. doi:10.1111/j.1467-7717.2009.01105.x

Samal KC (2003) Coping strategy of weaker sections (dalits and women) in post super-cyclone Orissa (case study of Ersamma block). Nabo Krishna Centre for Development Studies Publication, Bhubaneswar

Schoemaker P (1993) Multiple scenario development: its conceptual and behavioral foundation. Strateg Manag 14(3):193–213. doi:10.1002/smj.4250140304

Schön DA (1983) The reflective practitioner: how professionals think in action. Ashgate Publishing Ltd, London

Schwartz P (1996) The art of the long view: planning for the future in an uncertain world. Doubleday, New York

Sen A (1999) Health in development. Bull World Health Organ 77(8):619–623. http://cdrwww.who.int/bulletin/archives/77(8)619.pdf. Accessed 02 June 2017

Sen AK (2003) Forward. In: Farmer P (ed) Pathologies of power: health, human rights, and the new war on the poor. University of California Press, California

Sen AK (2009) The idea of justice. Penguin Books, London

Senge P (1990) The leader's new work: building learning organizations. Sloan Manag Rev 32 (1990):7–23

Senge PM, Kleiner A, Roberts C, Ross RB, Smith BJ (1994) The fifth discipline fieldbook. Doubleday, New York

Srivastava SK (2009) Making a technological choice for disaster management and poverty alleviation. Disasters 33(1):58–81. doi:10.1111/j.1467-7717.2008.01062.x

Sterman JD (2000) Business dynamics: systems thinking and modelling for a complex world. McGraw-Hill Publishing, Boston

Taleb NN (2007) The black swan: the impact of the highly improbable. Penguin Books Ltd, London

The Royal Society (1992) Risk: analysis, perception and management. The Royal Society, London

UN (2005) Hyogo framework for action 2005–2015: building the resilience of nations and communities to disasters. In: World conference on disaster reduction, Kobe, Hyogo, Japan, 18–22 Jan 2005

UN (2015) Sendai framework for disaster risk reduction 2015–2030. UNISDR, Geneva. http://www.unisdr.org/we/coordinate/sendai-framework. Accessed 24 June 2016

UNDP (2007) Human development report 2007/08. Human solidarity in a divided world. Palgrave Macmillan, New York, Fighting Climate Change

UNISDR (2015) Japan provides early warning example. In: UN world conference in disaster risk reduction 2015 Sendai Japan. Third UN World conference on disaster risk reduction. www.wcdrr.org. Accessed 16 March 2015

UPSC (2015) Civil services. UPSC Guide.com. www.upscguide.com. Accessed 12 May 2015

VCLL (2013) Practising reflection: reflective practises in disaster risk reduction: Short course handbook. University of Leicester, Leicester

Vickers G (1983) Human systems are different. J Appl Syst Anal 10(1983):3–13

Wahlström M (2015) From agreement to action: what next for the new global framework on disasters?. The Policy Lab Meeting at the Royal Society, London, 24 June

Weick KE (1988) Enacted sensemaking in crisis situations. J Manag Stud 25(4):305–317. doi:10.1111/j.1467-6486.1988.tb00039.x

White D (1995) Application of systems thinking to risk management. Manag Decis 33(10):35–45. doi:10.1108/EUM0000000003918

WHO (2002) Gender and health in natural disasters. WHO: gender, equity, human rights. http://www.who.int/gender/other_health/en/genderdisasters.pdf. Accessed 9 Dec 2014

Yin RK (2009) Case study research: design and methods, 4th edn. Sage, London

Yin RK (2012) Applications of case study research, 3rd edn. Sage, London

Index

N.S. Ray-Bennett *Avoidable Deaths*, SpringerBriefs in Environmental Science,
https://doi.org/10.1007/978-3-319-66951-9

MIX
Papier aus verantwortungsvollen Quellen
Paper from responsible sources
FSC® C105338

If you have any concerns about our products,
you can contact us on
ProductSafety@springernature.com

In case Publisher is established outside the EU,
the EU authorized representative is:
Springer Nature Customer Service Center GmbH
Europaplatz 3, 69115 Heidelberg, Germany

Printed by Libri Plureos GmbH
in Hamburg, Germany